D1713337

Housing in Postwar Canada: Demographic Change, Household Formation, and Housing Demand

Between 1945 and 1981 the Canadian population doubled, while the number of dwelling units more than tripled. John Miron shows how changes in demographic structure and housing affordability affected postwar household formation and housing demand. A central theme of his argument is that no single explanation adequately reflects the extent of what happened: a number of demographic trends were important, as were economic changes.

The size of Canadian households has been declining since at least the 1880s. Miron compares this trend to patterns of household size in Britain and the United States and argues that postwar changes in household formation in Canada were the result of several forces, including the postwar baby boom, increased longevity, changes in marriage patterns, rising incidence of divorce, increased household affluence, and new forms of government assistance in housing.

While aggregate growth in population, families, and households helps to explain why more housing was necessary, it does not explain changes in the kind of houses desired. Miron discusses changes in available housing stock as well as changes in structural type, such as the great apartment boom of the late 1960s and the re-emergence of owner occupancy in the late 1970s. The types of data available for measuring change in the stock and sources of error in housing data are also analysed. One of the book's most important contributions is an annotated synthesis of national trends in household formation and housing demand, derived from Statistics Canada's census data and accompanied by an insightful analysis of the relation of these trends in housing stock evolution. This is the only available detailed study of these topics in the Canadian context.

John R. Miron is professor of geography and planning, Scarborough College, and a research associate of the Centre for Urban and Community Studies, University of Toronto.

Housing in Postwar Canada

Demographic Change, Household Formation, and Housing Demand

JOHN R. MIRON

McGill-Queen's University Press
Kingston and Montreal

© McGill-Queen's University Press 1988
ISBN 0-7735-0614-4

Legal deposit first quarter 1988
Bibliothèque nationale du Québec

Printed in Canada

This book has been published with the help of a grant from the Social Science Federation of Canada, using funds provided by the Social Sciences and Humanities Research Council of Canada.

Printed on acid-free paper

Canadian Cataloguing in Publication Data

Miron, John R., 1947-
Housing in postwar Canada
Includes index.
ISBN 0-7735-0614-4
1. Housing - Canada. 2. Households - Canada.
3. Canada - Population. I. Title.
HD7305.A3M57 1988 363.5'0971 C87-093262-4

HD
7305
.A3
M57
1988

WITHDRAWN

JUN 20 1988

MGA

To my parents, and the other parents of the baby boom, who are in large part the subject of this book. Their hard work in rebuilding from the Depression, their optimism about the future world they would bequeath to their children, and their sense of fairness, which led to the "Just Society" and the modern welfare state, have left unmistakeable imprints on Canadian society in general and the baby boomers in particular.

To my gentle wife, Suzanne, without whose love, affection, and support this book could not have been written.

To my children, Peter and Marc, whom I want dearly to understand how our society has evolved and changed. It will be up to them and their peers to take on the challenges and to recognize where progress has been made and where problems remain. I hope that, in its small way, this book will help.

Contents

Acknowledgments

I began work on this book about four years ago, although it seems much longer in retrospect. My research in this area began a few years before that, in 1978. At the time, I was interested in how demographic shifts shaped metropolitan spatial form and growth. The immediate impetus for my own work was Gottlieb's careful analysis of long swings in urban development and Hickman's study of the cyclical nature of U.S. housing demand.

I have been very fortunate to have had a teaching appointment in geography at Scarborough College, University of Toronto. The college, indeed the entire university, puts a major emphasis on research. Over the years, the college has fostered my research activities and provided moral, intellectual, and financial support. I am very grateful for this attitude and atmosphere.

I have also been fortunate to have had a close relationship with the Centre for Urban and Community Studies at the University of Toronto. Since 1970, the centre has provided assistance in my research. The centre put up the initial seed money for my work on household formation in Toronto. The centre is a unique facility, both in terms of its interdisciplinary staff and its connections across and outside the university. Without the centre, it might not have been possible to get that all-important early outside sponsorship or valuable interdisciplinary perspectives and feedback. In addition, I owe special thanks to the staff of the centre, who have assisted me in ways too numerous to list.

Over the years, a number of other sponsors have provided financial support. These include the Ontario Ministry of Housing, Canada Mortgage and Housing Corporation (CMHC), the Social Sciences and Humanities Research Council of Canada, and the Ontario Commission of Inquiry into Residential Tenancies. The University of Toronto provided important financial support through a sabbatical leave in 1982–3 during which the book was initiated. Without such active support and interest, the book simply could not have been written.

I owe many thanks to my colleagues. At Scarborough College, they include the former division chairman, Rorke Bryan, who made sure that my research program was not sidetracked by the day-to-day problems and demands of university life. A similar function has been carried out at the centre by directors Larry Bourne and Meyer Brownstone. More than any one other person, Larry has been a source of encouragement and constructive criticism. Other research associates of the centre from whose presence I (and the book) have benefited include Barry Cullingworth, Marion Steele, and Barry Wellman. Outside the University of Toronto, I owe a special debt to Eric Moore of Queen's University. Over the years, he has had an ongoing interest in my work, and his insightful comments and support have been greatly appreciated.

A number of other persons have also read and commented on the monograph in its draft form. These include Sylvia Wargon and Janet Che-Alford of Statistics Canada, Philip Brown of CMHC, Pat Streich, formerly of Queen's University, and Rod Beaujot of the University of Western Ontario. Without meaning to offend anyone whom I have forgotten, I very much appreciate the suggestions that many people have made in regard to errors of fact or style in earlier drafts. Of course, I am solely responsible for any remaining errors.

I wish also to acknowledge the efforts of a small army of research assistants who have given some of the best years of their youth to this work. They include Sharon Smith, Gary Bryant, Larry Lall, Pam Parsons, Pamela Blais, Linda Geall, Peter Dunn, Laurien Brunet, Mary Jane Braide, and Lucia Lo. Their enthusiasm, keenness, and hard work made some apparently insurmountable research problems look easy.

Housing in Postwar Canada

CHAPTER ONE

Introduction

After the Second World War, Canada entered a new period of prosperity and rapid social change: an experience shared by several other developed countries. In the interwar period (1918–39), population growth rates had fallen precipitously, and projections abounded of absolute population decline in the 1950s. Doubts were commonplace about the ability of the postwar economy to recover from the dismal Depression of the 1930s, a prospect further enhanced by the spectre of population decline. However, these sombre predictions never materialized. Between 1946 and 1981, Canada's population doubled. At the same time, there were remarkable shifts in its geographical and social structure. Rapid urbanization, the baby boom and subsequent bust, increasing longevity, and changing marriage patterns were among these. They were accompanied by surging incomes that outpaced consumer prices in general and housing prices in particular. The two sets of forces – demographic and economic – led to changes in the living arrangements of individuals and their housing.

And the changes were substantial. To get a feel for the extent of change, let us compare Census data for 1941 and 1981. There was no census in 1946. However, given that the housing stock did not change much during the war, 1941 data are probably not wide of the mark. It should not be surprising that a burgeoning population meant more housing. The uninitiated would expect the housing stock to keep pace with population growth. However, while population doubled, the number of dwelling units more than tripled. Suppose we choose at random enough households to constitute a population of 10,000. In 1941, those households would have occupied 2,240 dwellings. In 1981, the same population filled 3,480 units. During the same period, there was a remarkable geographic shift. Of those 2,240 typical dwellings in 1941, only about half were in urban areas. By 1981, urban areas accounted for almost 80 per cent of the housing stock. See Table 1.

TABLE 1
Population, households, and dwellings by geographic locale, Canada, 1941 and 1981

	All Canada*	Rural		Urban
		Farm	Nonfarm	
*1941**				
Population (000s)†	11,490	3,279	2,358	5,852
Occupied dwellings (000s)‡	2,573	704	556	1,313
Average persons per dwelling	4.5	4.7	4.2	4.5
Dwellings per 10,000 persons§	2,240	613	484	1,143
Dwellings by tenure				
% rented, other#	43	19	34	60
% owner-occupied	57	81	66	40
Dwellings by type				
% single detached	71	97	87	49
% semi-detached, row	8	1	5	12
% apartments, flats, other#	22	1	9	38
*1981**				
Population†	23,797	1,014	4,775	18,009
Occupied dwellings‡	8,281	271	1,505	6,506
Average persons per dwelling	2.9	3.7	3.2	2.8
Dwellings per 10,000 persons§	3,480	114	632	2,734
Dwellings by tenure				
% rented	38	5	18	44
% owner-occupied	62	95	82	56
Dwellings by type				
% single-detached	57	95	83	50
% semi-detached, row	10	1	3	11
% apartments, flats, other	33	5	14	39

SOURCE: Censuses of Canada, 1941, 1981
* 1941 data exclude Yukon and the Northwest Territories. 1981 data include Newfoundland.
† Total population used. Includes individuals in collective dwellings
‡ Only private dwellings counted. Collective dwellings omitted
§ Shows national breakdown. Thus 10,000 persons nationally included 114 farm dwellings, 632 rural nonfarm dwellings, and 2,734 urban dwellings in 1981.
# In 1941, "other" dwellings by type or tenure include a small number of cases where type or tenure was not recorded.

Between 1941 and 1981, the character of the housing stock changed substantially. Overall, there was a small increase in the incidence of homeownership. However, this belies an important change. Traditionally, rural households in Canada had been predominantly owners, urban households mainly renters. With the shift to urban areas, the incidence of homeownership overall might well have declined. This was counteracted, though, by a sharp rise in the incidence of urban homeownership, from 40 per cent of households in 1941 to 56 per cent in 1981.

Also remarkable was the changing composition of the housing stock.

In 1941, over 70 per cent the stock consisted of single detached dwellings. Although there has always been a varied mix of new housing, the predominance of particular building forms did change quickly. In the immediate postwar building boom, the smaller (typically one and one-half storey) detached house prevailed. This gave way, in the early 1950s, to larger bungalows, split levels, and then two-storey houses. The period from the early 1960s to the 1970s also saw the great apartment boom, in both inner-city and suburban areas. The legalization of the condominium entity, in the late 1960s, led to extensive construction of highrise and townhouse units of this type. Finally, extensive construction of assisted (rentership and ownership) housing after about the mid-1960s should also be noted. By 1981, less than 60 per cent of the housing stock was single detached. New single detached units accounted for only about one-half of all postwar housing construction. The large apartment building emerged as the new housing form of the postwar period. By 1981, apartments and flats made up one-third of the housing stock.

This book is concerned with the reasons for these changes in household formation and the housing stock. Why did they occur? To what extent were they a reflection of the changing demographic characteristics of Canada's population? To what extent did they reflect postwar prosperity and the changing price of housing? And to what extent were household formation and housing stock change interdependent?

HOUSEHOLD FORMATION AND HOUSING DEMAND

This book is about how people organize themselves into households and the dwellings in which they live. These notions are defined more specifically later in the book, but for the moment let us use the following. A "dwelling" is a set of living quarters. It might, for example, be an apartment in a large building; a flat in a house; one unit in a duplex, triplex, or other small apartment building; a row, semi-detached, or detached house; a mobile home; a houseboat; or a tent. A "household" is simply the collection of people "usually resident" in that dwelling. A household might consist of a person living alone; it might be a family; it might be some combination of one or more families and/or unrelated persons.

I refer to the patterns in which individuals form their living arrangements as "household formation." Throughout the book, household formation is equated with "choice of living arrangement." From among feasible alternatives, people choose according to their preferences and the prices (or affordability) of the alternatives. Thus choice reflects the preferences and income of the consumer and the prices and availability of alternatives. In an economic perspective, the problem of choice is the stuff of demand analysis. This book adopts such a perspective in interpreting postwar household formation. In this view, demographic changes reshaped consumer preferences, and the interaction of these with income and price change produced new choices of living arrangements.

Similarly, the housing occupied by a household may be viewed using demand analysis. The choice of tenure, size, or type of dwelling can be cast in terms of income, preferences, and the prices of available alternatives. This allows us to look at postwar housing stock change in terms of the demographic forces that shaped consumer preferences as well as income and housing prices (i.e. housing affordability).

For the most part, this book uses demand analysis to look at household formation and housing stock change. However, in any market, there are both demand and supply forces. To what extent did changes in the supply curve of housing also affect choices of living arrangement and housing stock change? Although a fascinating question, it is largely beyond the scope of this book. To answer it would require detailed analysis of the housing supply sector in terms of changing technology and input prices. In this book, supply aspects are given only limited attention: discussed are changes in postwar building technology and the impact of public-sector policies on the housing stock choices available to consumers. It is assumed simply that supply forces are reflected in the price of housing, and this study examines the response of consumer demand to this. In other words, if technology changed in a way that made house production less expensive, it should be reflected in lower housing prices, hence increased demand.

Finally, the meaning of "housing" itself needs to be clarified. This book is based largely on Census data. It considers the one unit of housing wherein a person is usually resident on Census day. It ignores the fact that, over a period of time, individuals may regularly live in several different dwellings and living arrangements. These might include a dormitory or work camp, vacation or second home, motor or mobile home, vacation time-sharing unit, or hotel room. These form part of one's housing demand, but there is little information available about them. Unfortunately, the substitution between different forms of accommodation may well have been important. To what extent, for example, did metropolitan households compensate for higher land prices by choosing more land-intensive housing (i.e. a smaller lot or greater use of semi-detached, row, or apartment housing) while also consuming more land-extensive vacation homes? Although this book cannot answer such questions, it will be shown that such considerations were potentially important.

EXPLAINING POSTWAR CHANGES

There are almost as many explanations of postwar change in household formation and the housing stock as there are observers. These explanations are not necessarily incompatible or mutually exclusive. A central argument of this book as that we must be careful not to assume that only one is correct or important. What happened in the postwar period was the result of several forces acting in concert. To get an idea of where the book is headed, let us consider a list of explanations.

The postwar baby boom. From 1946 to 1960, a sustained high rate of fertility was accentuated by the prewar fertility slump and another slump that began in the early 1960s. As the "baby boom" generation grew up, several related changes in household composition affected housing demand. In the postwar baby boom years and immediately thereafter, there were many young, large families. These families fuelled the postwar suburban development boom, leaving an unmistakeable imprint in terms of low-density, family-oriented housing. Detached housing with large yards for playing, more bedrooms, and basement "rec rooms" are artifacts of this period. With the subsequent drop in fertility, young families had fewer children than before. The emphasis in housing demand shifted correspondingly: e.g. less concern with yards and number of bedrooms and greater emphasis on special-purpose rooms. Further, the parents of the baby boom then became empty nesters, and households shrank with the departure of grown-up children. Although many empty nesters continued to live on in a larger, family-oriented home, their housing needs and preferences changed.

Changing life cycles. A life cycle can be described for many individuals demarcated by certain major events: marriage, birth of the first child, birth of the last child, and the home leaving of the last child. Today, as in times past, most individuals experience all these events. However, there have been changes in the typical ages at which these major events occur. Comparing 1980 with 1970, for instance, a typical women married and had her first child later. However, she tended to have here last child earlier and the last child was expected to leave home earlier. These life cycle changes indicate a compression of the child-rearing phase of family life. In other words, the proportion of a married couple's lifetime spent with children around the home declined. When combined with the shrinking completed family sizes, this affected typical household size and composition. This life cycle shift also meant that child-rearing considerations became less important in housing demand.

Increased longevity. Throughout the twentieth century, life expectancies increased. In part, declining infant mortality meant a greater chance of reaching old age. In part, however, remaining life expectancy also increased for the elderly. The increasingly large gap in life expectancy between the sexes was also noteworthy. Increased longevity raised the chance of a couple surviving intact to empty-nesterhood and thereby the frequency of two-person households. The increasing sex differential in mortality meant that more women outlived their husbands, and for longer periods of time. To the extent that widows live alone, rather than with other family members or friends, the number of one-person households increased. Some of this aging population continued to live on in family dwellings. Others found such houses too large for their needs, or too difficult to maintain, and moved, creating a demand for smaller or otherwise more suitable quarters.

Marriage rush and marriage bust. In the postwar period, marriage went through a phase of popularity, then decline. In the early postwar period, the so-called marriage rush, adults of all ages were more likely to marry. They tended to marry younger and had a high probability of ever marrying. Then, beginning in the 1960s, the popularity of marriage waned. The post-1960 period is sometimes called the "marriage bust." It was unclear at the time of writing whether the decline in nuptiality was simply a delay in marriage. In part, it may also reflect the popularity of alternatives to marriage, such as cohabitation. Whatever the reason, the effect on household formation was marked. Persons who chose not to marry were more likely to live alone (i.e. in one-person households), although some lived with parents or friends or cohabited. Such persons also differ typically from families in terms of their dwelling choices.

Earlier home leaving. In many societies, young adults leave a parental home at some stage to establish their own households. Often this is upon marriage. In the postwar period, it became increasingly common for young Canadians to leave home. Sometimes they left to go to college, sometimes to take a job, and sometimes to have more privacy or freedom from parental influence. Such home leaving had implications for average household size. It decreased the size of the parental household. There were a variety of possible living arrangements for the home leaver - from lodging to sharing space to living alone - each with particular implications for household formation and housing demand.

The rising incidence of divorce. In Canada, divorce law was reformed in 1968. Under the new law, divorce procedures were simplified and the grounds for divorce extended. Subsequently, there was a rapid increase in the incidence of divorce. Household formation or reformation is an integral part of divorce. Typically, the wife retained custody of any children and what was formerly one household split into two, the second often being the husband living alone. This resulted in smaller households. It sometimes also resulted in two households living off an income that formerly supported one: with consequent implications for housing demand.

The demand for privacy. In the postwar period, there was a considerable "undoubling" in accommodation. Much of this, particularly in the 1950s, was an undoubling of families. Households of two or more families became rare. Undoubling also includes the declining incidence of persons other than parents and children in family households. This latter type of undoubling was especially important in the 1960s and 1970s. These two kinds of undoubling accounted for some of the decline in average household size. It may also have accounted, as a consequence, for changes in the demand for large-dwellings. One view of this undoubling is that it simply reflected a heightened desire for privacy. Families and individuals undoubled because they increasingly came to value living alone (i.e. privacy).

Housing affordability. While preference for a living arrangement (and the privacy it engenders) is important, so too is the affordability of that living arrangement. The consumer must take into account the cost of a living arrangement relative to income and the prices of other goods. The postwar period was characterized by a substantial improvement in household incomes. At the same time, the price of housing increased only modestly relative to both incomes and the prices of other goods. In part, individuals used this increasing affordability to form households and to consume particular kinds of housing.

The role of the public sector. For much of the postwar period, governments at the federal, provincial, and local level attempted directly to regulate or stimulate the stock of housing. These activities also affected household formation in a variety of, at times offsetting, ways. For one, they set minimum requirements that made new housing more expensive and thus affected choices of living arrangement. For another, they provided a variety of assistance programs for those who could not otherwise afford decent accommodation. These housing programs broke the link between income and affordability. They enabled the formation of households that might otherwise not have been able to afford separate accommodation and made particular types of accommodation more widely affordable to existing households.

Improvements in home-making technology. Having a separate dwelling represents chores as well as freedom. Typically, food has to be prepared, rooms and laundry have to be cleaned, and maintenance work has to be done. These tasks deter some individuals and families from establishing their own dwelling. However, remarkable changes in home-making technology in the postwar period made many of these chores easier. Electric dishwashers, microwave ovens, slow cookers, frost-free refrigerators, vacuum cleaners, stain repellants, and permanent press clothing were some of these innovations. By making home-making easier, they contributed to the number of individuals who found it feasible to live alone or in a small household and stimulated the demand for smaller dwellings.

MEANING OF "EXPLANATION"

Some observers argue that there is a single, simple explanation for postwar changes in household formation and the housing stock. For instance, some see the changes simply as a blip resulting from the baby boom. The above list is a reminder of the varied processes that have affected the organization of individuals into households and dwellings. Evidence presented in this book suggests that, while some of these are relatively more important than others, a variety of explanations is needed to account for much of the observed change.

Let us note a caveat at this point. The above conclusion is based on numerical estimates of the relative importance of each explanation in accounting for observed changes in household formation and the housing stock. Empirical estimates in turn are based on theoretical constructs. Developing such estimates is worthwhile in part because it makes us realize what we do or do not understand.

Further, one must always be careful in assessing the importance of an "explanation." In a sense, there can never be "an explanation," only a "better explanation." The decisions of individuals to form a household or to live in a certain dwelling are the stuff of life itself. They can involve complex decisions of the heart. We can (and will) say, for example, that the increasing incidence of divorce led to the formation of "x" more households. However, it is another thing altogether to explain why the incidence of divorce has been rising. The objective of this book is to estimate empirically the effects of changes in demographic structure (e.g. age mix, marital propensities, fertility, and mortality), housing policy, and housing affordability on postwar household formation and housing demand. It does not extensively explore what lies behind these changes.

As in the case of the "social" explanations, empirical estimates of the affordability explanation need to be carefully qualified. We can (and will) assert that rising real incomes led to more household formation among young marrieds, for example. However, this is only a part of the effect of growing prosperity. Some people argue, for instance, that divorce rates rose in part because of growing affluence. As they became better off, couples found that the monetary costs of separation and divorce became more affordable. If so, increasing incomes have caused related changes in household formation. Such issues are largely ignored here, but the reader should be mindful of them in evaluating the empirical estimates presented below.

Finally, let us clarify the sense in which the term *explanation* is used. A certain argument can be a "valid" explanation to one person but meaningless to another. An economist, for example, might conclude that an observed change in housing demand is attributable to a "change in taste." To another economist, this might be valid; it means that a rational consumer, faced with the same prices and income, would have made a different choice, based on past observation, if his tastes had not changed. To a sociologist, however, this might be no explanation at all. As Beresford and Rivlin (1966, 254–5) state: "A 'change in taste' is the economist's phrase for something he is unable to explain." Equally, a sociologist's explanation might be criticized for ignoring the role of prices. To different people, any of the above explanations may raise as many questions as it answers. That is not a fault. It is in the nature of science to continue questioning and expanding upon explanations.

ISSUES OF RELEVANCE TO PUBLIC POLICY

An individual comes to live in a particular household and dwelling because of a sequence of circumstances and decisions that have to do with his or her aspirations in life, friendships, loves, family ties, income, and the prices and availabilities of housing alternatives. To a social scientist, this is reason enough for being interested in how postwar patterns of household formation and housing demand were determined.

However, a better understanding of what happened is also relevant to contemporary public policy. Several current issues of public concern have been or are fundamentally affected by changes in household formation and the housing stock. For planners, housing market analysts, and others, it is often these public concerns that motivate their interest in this topic. Let us briefly consider a selection of public concerns prevalent in the mid-1980s.

The Greying of Postwar Suburbia

The greying of postwar suburbia has become a pressing issue in some large metropolitan areas. From 1945 to the mid-1960s, vast tracts of land were developed in Canada and the United States to suit the housing needs of the young, large families that sprang up with the postwar baby boom. Local governments built schools, playgrounds, children's libraries, community arenas, and other facilities to meet the needs of such households.

In later years, these children grew up and left home. The parents, left behind, are commonly referred to as "empty nesters." Some of these empty nesters moved to smaller accommodation or retired and moved away from the metropolitan area altogether. However, many others continued to live on in the same dwellings in which they raised their families. It is these latter non-movers who are the focus of several public policy concerns. See, for example, Myers (1978).

The Declining Population of the Central City

Consider the case of the city of Toronto, a central city area that includes about one-quarter of the population of Metropolitan Toronto. The city was virtually completely built up before the Second World War; there was effectively no postwar suburban development within its boundaries. Through careful planning and good fortune, the city was able to retain viable residential neighbourhoods and a mixed local employment base. It did not suffer from housing stock neglect or abandonment common in other large North American cities. Although there was extensive commercial and office development, the city was able to maintain its overall stock of dwellings. In spite of this, however, its population fell by about one-seventh between 1971 and 1981.[1] With declining average household sizes, the same housing stock accommodated a smaller population over time. This has chiefly been because of the rise of the one-person household, which made up, for example, over one-third of all households in the city of Toronto in 1981. In some city census tracts, over two-thirds of all households contained just one person. This experience is not unlike that of other large Canadian cities.

Future Housing Demand

How much housing will be needed over the next few decades? Can these demands be met from the existing housing stock, or is substantial new construction required? Will new public policy initiatives be needed?

Some analysts argue that the housing construction industry in the early 1980s underwent a major structural change. They believe that future volumes of housing construction will be much lower than those observed in the postwar period. It is argued that the postwar building boom was fuelled at first by a need to upgrade the existing housing stock to satisfy the more affluent and larger "baby boom" families and later by the emerging housing needs of the aging baby boomers themselves. No new source of housing demand has emerged, and it is argued that housing construction will taper off to replacement levels after the mid-1980s as the last of the baby boom children form their own households.

Other analysts argue differently. They recognize the role of the maturing baby boom but argue that new construction need not fall all the way back to replacement levels after these baby boomers finish entering adulthood. They argue that other kinds of demographic change - in life expectancy, marital propensities, age at home leaving, and the tendency to share accommodation - will continue to mean substantial new housing construction.

Most analysts do not foresee a major change in fertility; small families are almost universally expected. However, they disagree about the nature of their housing demands. In part, this is because of uncertainty about how future incomes will compare to housing prices. In part, there is also uncertainty about how these incomes might be spent. If the future small family is able to afford better housing, will it seek larger or more luxurious housing than is typically demanded by such households today?

The future of new housing construction is of considerable interest to the private construction sector. It is also of interest to housing policy analysts who want to know if more assisted housing is needed. Also, since construction activity is closely watched as a general economic indicator, it is important to know if there is a fundamental change under way in patterns of housing demand.

New Housing Markets

Related to this is an argument about the emergence of new housing markets. Some observers, while recognizing that the aging of the baby boom means a reduction in the housing industry's traditional source of demand, are still optimistic about market prospects. They argue that new housing demands will appear to replace those of the dwindling number of young families.

Where are these new demands to come from? Usually it is argued that growing affluence, combined with changing life-styles, will lead to new "mature market" demands. One oft-cited example is empty nesters. It has been suggested that there is a large potential demand for new housing aimed at affluent empty nesters. Others argue that the smaller family household of the future will have more income to spend on housing and thus will want more luxurious homes: homes that, while not necessarily physically larger, are better suited to their life-styles than what is currently available.

Are such markets likely to emerge? Can we forecast anything about the future numbers and housing demands of empty nesters, small families, or other target groups? Answers to such questions are valuable to the private sector and to public policy analysts.

DATA SOURCES

What information would help us better to understand the choice of living arrangements of these people and the kinds of dwelling units they occupy? We need to know something about the characteristics of the individual. Presumably, we are interested in family status, because the kind of household lived in will likely depend on whether the individual is a family member (i.e. a spouse, parent, or child). Family status also tends to affect the kind of dwelling occupied, as families have traditionally been more likely to be owners than have non-family persons. We might also want to know about income, wealth, and job history. Such data provide an idea of how the choice of living arrangement and dwelling was constrained by affordability. Some information on past residential history - where one has lived, how frequently one has moved, and recent changes in household composition (e.g. gain or loss of room-mates or lodgers, death or departure of spouse, and addition, loss, or departure of children) - could also help explain the current living arrangement and dwelling choice. We might want also to know about other members of the current household: e.g. number of people living there, family relationships, ages, incomes, wealth, and job histories. Such information could help explain how the living arrangement was formed. Further, we might want to know the characteristics of the dwelling. These include size (e.g. number of rooms, floor area), type (e.g. detached dwelling, townhouse, flat), tenure (e.g. owner-occupied, rented), cost, and (if applicable) down payment. Given similar information for other dwellings in the same local housing market, we can examine the dwelling choice tradeoffs facing a household.

While no one available data source has this range of detail over the entire 1941-81 period, the best sources are censuses. They detail the composition of households and their dwellings. Such data are used extensively throughout this book. However, censuses have little information on wealth, past residential history, or the cost of acquiring and operating the chosen dwelling or rejected alternatives. Also, it is difficult to obtain detailed information from censuses prior to 1971. Although census questionnaires have been retained, they are not publicly accessible. As a result, pre-1971 census information is largely limited to published summary tables. While valuable in themselves, they are often not detailed enough to permit elaborate analysis.

However, beginning with the 1971 Census public use samples, access to detailed census data greatly improved. The public use samples are 1 per cent random samples of census returns, suitably coded to protect confidentiality. In effect, the researcher is given a miniature version of the Census with which to create customized summary reports, at whatever level of detail. This book makes considerable use of these samples.

Government agencies such as Statistics Canada began, in the 1970s, to make available, in addition to the public use samples, other survey sample data on a similar basis. These agencies had, for many years, collected data for large samples of individuals and households. Until the 1970s, researchers outside government could obtain such data only from summary reports. However, beginning in the early 1970s, these agencies began releasing micro data samples that gave individual or household responses to many of the survey questions. Some data were omitted or altered to guarantee confidentiality, but, broadly speaking, the researcher was given a large data file that looked much like the original survey data.

However, even with all these sample data, problems remain. One common problem is that the observational unit tends to be the household or the "head" of household, rather than the individual. We thus typically know much about the household head but relatively little about the other individuals who live with him or her. In such cases, we have to use the head's age, sex, and marital status and household size, presence of children, and family composition data to infer the characteristics of the remaining household members. In a similar way, we often have to guess at the household's past residential history and its wealth. Although later censuses were computerized to permit such tabulations, little comparable information is available from earlier censuses.

Another problem is that these samples never include data on the relative costs of housing alternatives. Typically, one must rely on cost data for otherwise-similar households that have made different dwelling choices. This presupposes that we can identify households in similar circumstances. Although not entirely avoidable, this problem can be minimized by careful, detailed delineation of cohorts. A cohort is simply a group of individuals or households that share a certain set of characteristics. Ideally, each cohort would consist of "similar" individuals or households. However, given that we want to distinguish between high and low incomes, between larger and smaller households, between large-city residents and those in small towns, and so on, the number of cohorts needed often is very large. There is little point in defining a large number of cohorts if our sample contains no, or few, households or individuals in many of them. The sample has to be large enough to permit reliable estimation for each cohort designated. In practice, this means often that the sample has to be of a size found only in a census.

This book also makes use of data and statistical findings for the United

States and other countries over the same period. In many respects, the American experience in household formation was similar to that in Canada. Looking at U.S. data helps provide information where similar detail is not available for Canada. At the same time, some comparative differences in experience are highlighted that help in understanding why the living arrangements and housing choices of Canadians changed as they did.

OUTLINE OF THE BOOK

The remainder of this book can be divided broadly into three parts. One part includes chapters two and three. Chapter two discusses the definitions and concepts that underlie much of the rest of the book. This includes definitions of family, household, and dwelling. Also considered are indicators of dwelling size and quality and household composition. Chapter three traces postwar patterns of family formation in Canada and relates these to demographic changes. It reviews the changes in fertility, mortality, migration, and nuptiality that have affected family formation and size distributions. These two chapters provide a foundation for the remaining two parts.

The second part of the book analyses the process of household formation. In chapter four, postwar changes in household formation are described and related to changes in family formation. The importance of nonfamily individuals in postwar household formation is highlighted. Chapter five describes a "components of change" analysis of household formation. Using this technique, I develop and compare numerical estimates of the relative importance of alternative explanations of household formation over the 1970s. Chapter six considers the effect of rising incomes on household formation. Using cross-sectional data, I show families and nonfamily individuals in general to be more likely to head a dwelling, and more likely to live alone, the higher their income. The effect of rising incomes over time on household formation is estimated. Conclusions are drawn about the relative importance of economic and demographic explanations of postwar household formation.

The final part of the book looks at housing stock change and its interrelationship with postwar household formation. In chapter seven, postwar changes in Canada's housing stock are described. This includes a review of the changing mix of dwelling types, tenures, size, and quality. Chapter eight presents a discussion of the changing relation between housing demand and affordability over the postwar period. It assesses the role of income and house prices in determining housing consumption and contrasts this with demographic forces such as changing household size. This permits an interpretation of the effect of household formation on housing stock change. Chapter nine reviews the substantial role of the public sector in shaping the postwar housing stock. It discusses the wide variety of explicit and implicit subsidies provided to owner-occupiers and rental housing developers, at both the federal and local level. It considers also the effects of regulatory constraints such as land subdivision controls, zoning bylaws, and building codes. The potential effects of such regulation and subsidization on the rate and pattern of household formation are discussed.

Chapter ten summarizes the conclusions reached in this book about the impact of demographic change on household formation and housing demand in postwar Canada.

CHAPTER TWO

Definitions and Concepts

What exactly is a dwelling, a household, or a family? In chapter one, some tentative definitions are given. In this chapter, these and other definitions are expanded and made formal.

That this book uses data from census and other large micro data samples largely constrains the definitions employed. These data were collected for a number of purposes, analysis of household formation and housing demand being but one. The kinds of data collected and the concepts and definitions employed have to serve a variety of interests and therefore may not be the best from the point of view of this book. To their credit, however, these various micro data samples have employed similar definitions, thus permitting empirical comparisons between samples. Other merits of these definitions are that they are simple, relatively unambiguous, and amenable to measurement.

The definitions used in this book, as in most of the micro data samples, follow closely those of the census of Canada in the 1970s. There may be a tendency to accept census definitions unthinkingly if only because of the many data that are based on them. However, census definitions are not without their own peculiarities and traps for the unwary. These are explored in the ensuing discussion, and the limitations of these definitions and concepts in household formation and housing demand research are clarified.

A DWELLING

For the most part, we are a nation of home dwellers. Almost all of us have one place, a dwelling, that we call home, where we sleep, eat, and carry on other activities. It is an address where others can find us. It is a place where we typically feel that we are away from the rest of society, where we are most free to do as we please (subject, perhaps, to the interests of others living there).

In early censuses in Canada - those up to and including 1921 - a dwelling was defined by its exterior appearance.[1] An apartment building, for example, was treated as one dwelling: regardless of the number of apartments it contained. In subsequent censuses, a dwelling was defined by its internal usage: e.g. each apartment in an apartment building or each half of a semi-detached building was a separate dwelling. This latter concept has come to be widely used in defining a dwelling.

Even in censuses, however, it has not been possible to come up with a completely unambiguous definition. Canadian censuses have used the same definition of a dwelling since 1941. However, the 1951 Census reported that (vol 10, 351):

> There has been some tendency for certain living quarters to be counted as separate dwellings in the 1951 Census which would not have been counted as such under the 1941 definitions. In other words, part of the increase in dwellings [counted] between 1941 and 1951 has been due to definitional changes. The actual extent to which these definitional changes have influenced the count of dwellings is impossible to determine with certainty. There are indications that the outside figure for Canada would not exceed 170,000 with the true figure probably closer to 100,000. There are also good indications that these differences due to definitional changes are confined almost entirely to the apartment and flat type of dwelling.

What happened here? Although the published definition did not change, its interpretation had. In other words, the definition used was sufficiently ambiguous to permit substantially different interpretations.[2]

In this book, a "private" dwelling is formally defined, as in the Census, as follows: "a structurally separate set of living quarters with a private entrance from outside or from a common hall, lobby, vestibule or stairway inside the building. The entrance to the dwelling must be one that can be used without passing through the living quarters of someone else" (*1981 Census Dictionary*, 85). With some variations in interpretation, this same definition was employed in every census in Canada between 1941 and 1981.

There are two elements to this definition. First, the dwelling must be a "structurally separate set of living quarters." I have not seen a clear explanation of this phrase in published census reports but have been told by census officials that it means that the occupants of a dwelling do not share certain rooms with occupants of other dwellings. For example, if I live alone in a one-room flat but share a kitchen or bathroom with persons living in other flats in the building, I would not have a structurally separate set of living quarters. However, this does not apply to all rooms. Some large apartment buildings, for example, have common laundry rooms, games rooms, exercise rooms, and so on. Such rooms are shared, yet the census treats the individual apartments as separate dwellings. Thus the shared-use rule applies only to certain kinds of rooms.

Second, a dwelling must have a "private entrance." Living quarters that can be reached only by passing through someone else's living quarters are not considered separate dwellings. A concept of privacy is inextricably woven in to this definition. A dwelling is defined to be a physical unit within which basic life functions, such as eating, sleeping, or toilette can be carried on in the absence of persons who are not members of the household. Further, entry to a dwelling cannot require passage through some other dwelling, thereby helping to preserve the privacy of that dwelling.

The above definition is for a "private occupied" dwelling. In the census, other dwellings, termed "collective occupied" dwellings, are also recognized. A collective (occupied) dwelling is defined as follows: "a dwelling of a commercial, institutional, or communal nature . . . Included are hotels, motels, tourist homes, nursing homes, hospitals, staff residences, communal quarters of military camps, work camps, jails, missions, rooming or lodging houses, and so on" (*1981 Census Dictionary*, 82). Also counted as a collective is any dwelling containing 10 or more unrelated persons. Thus a collective dwelling is defined either by its physical structure or by the social relations among its inhabitants. This book is concerned almost exclusively with private dwellings. For parsimony, we omit the use of the adjective "private" except where collective dwellings are discussed.[3]

A definition is one thing; its implementation is another. The notion of a dwelling is crucial to most censuses because it helps census-takers to ensure that each person is counted exactly once – at his or her usual place of residence – and to be neither missed nor double-counted. In Canadian censuses, which have been based on self-enumeration since 1971, Statistics Canada identifies and enumerates each dwelling. Nowhere is there an accurate list, though, by which this might be done. Statistics Canada's problem is to find out whether there are one, two, or more dwellings in a given building. It uses a variety of data sources to attempt to identify all the dwellings in a given area and supplements these by having enumerators watch out for additional dwellings. While delivering questionnaires, enumerators inquire when they suspect still more dwellings at a given address. The questionnaires themselves also include questions designed to detect such cases. However, there may be little chance of catching certain cases such as basement or upstairs flats if the respondent chooses to evade these questions.

In concluding, let us draw a distinction between the set of dwellings used for most of this book and the stock of housing units in Canada. In the 1981 Census of Canada, the typology illustrated below was employed.

FIGURE 1

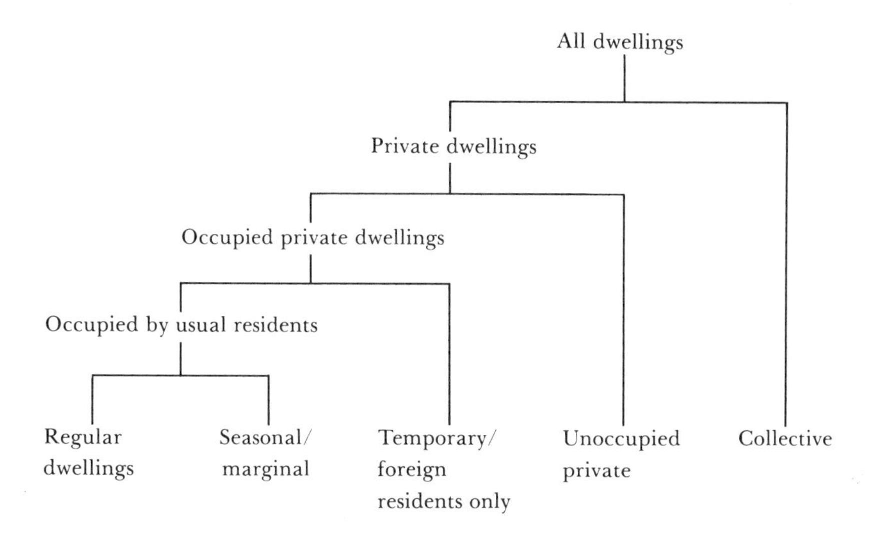

Seasonal/marginal dwellings are units that are unsuitable for year-round accommodation. Note, however, that some such dwellings can be readily and inexpensively "winterized," converted to year-round use. All other private dwellings are termed "Regular Dwellings." The Census distinguishes between occupied and unoccupied units. It further distinguishes between regular private dwellings occupied by usual residents and those occupied solely by foreign residents (i.e. persons whose usual place of residence is outside Canada) or temporary residents (i.e. persons whose usual place of residence is elsewhere in Canada). See Figure 1. In general, detailed information is available only for private dwellings occupied by usual residents. See chapter seven for data on the number of dwellings thus omitted.

Such a focus is broadly compatible with the interests of this book. We are interested in relations between household characteristics and features of the dwellings those households usually occupy; therefore a focus on occupied private dwellings is desirable. At the same time, this focus excludes some dwellings that make up part of the housing stock of Canada: i.e. collective dwellings, vacant dwellings, and dwellings occupied temporarily or by nonresidents.

SIZE AND QUALITY OF DWELLING

Later chapters examine the dwelling choices of households. Inevitably, this involves comparisons of the size or quality of different dwellings.

Most people have some sense of the size or quality of a particular dwelling. We might walk in, take a look around, and say "This is a large dwelling" or "This is well built." In making such statements, we usually combine many different bits of information. Using just one bit, or a few bits, of information such as the lot size, number of rooms, or floor area can be misleading because of the variability among such indicators from one dwelling to the next. However, taken together, as in the following real estate advertisement, the overall impression is unmistakeable: "12 room, five bath home, with swimming pool, and tennis courts . . . The master suite alone is an elegantly appointed indoor/outdoor complex of some two thousand square feet with a solar greenhouse sitting room equipped with fire place and wet bar, a master bath in Italian marble and 18k golden fixtures; its centerpiece, a sunken 6′ x 6′ Waterjet whirlpool bath, directly under a skylight; a cedar closeted, totally mirrored dressing room; sliding doors opening onto an expansive rooftop terrace. The restaurant sized kitchen and pantry can handle a formal dinner for twelve, or a buffet for hundreds. The circular driveway can accommodate 22 cars. And the guest suites are simply superb."

Unfortunately, housing analysts have few data on dwelling size or quality. Micro data files rarely contain photographs of the dwelling units, or even detailed summaries of size and quality information. Typically, they detail only the number of rooms, number of bedrooms, number of bathrooms, and type of dwelling (e.g. detached, mobile home, semi-detached, duplex, townhouse, or apartment/flat). Sometimes, rental value or estimated selling price is also presented. Much less frequently, one finds floor area and lot size. The analyst's problem is to construct an overall measure of dwelling size or quality from such data.

An additional problem is that even these few variables can be defined in different ways. Consider, for example, the number of rooms in a dwelling. In the 1971 and 1981 censuses, a room is defined as follows: "A room is an enclosed area within a dwelling which is finished and suitable for year round living . . . Partially divided L-shaped rooms are considered to be separate rooms if they are considered as such by the respondent (e.g., L-shaped dining room living room arrangements). Not counted as rooms are bathrooms, halls, vestibules, and rooms used solely for business purposes" (*1981 Census Dictionary*, 89). This excludes "unfinished" areas, such as in an attic or basement, without defining how much carpentry work is required to "finish" it. It excludes closets or hallways, although these may be used for purposes other than storage or passage. Finally, it allows for individual interpretation of what constitutes the dividing up of a given space into rooms. L-shaped rooms can be counted as one or two rooms, for example.

The floor plan in Figure 2 shows an "open plan" bungalow with broad

FIGURE 2

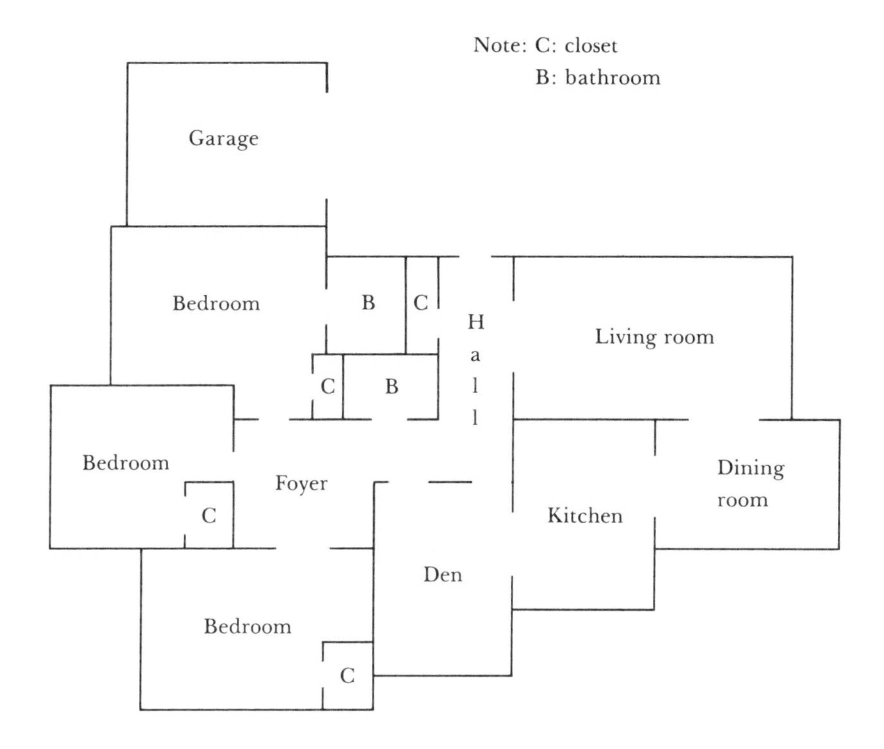

walkthroughs joining the living room to dining room to kitchen to den. Note also the foyer that provides entry to the three bedrooms and could serve as a telephone or conversation room or as a second den. The number of rooms counted could range from four to eight, depending on one's interpretation.

Earlier censuses each used similar, but not identical, definitions. "The following shall not be counted as rooms: halls, bathrooms, closets, pantries and alcoves; attics and basements, unless finished off for living purposes; sunrooms and verandas unless suitably enclosed for occupancy at all seasons" (*1941 Census of Canada*, vol 5, xvii). The *1951 Census of Canada* noted: "Only rooms used or suitable for living purposes, including those occupied by servants, lodgers, or members of lodging families, were counted as rooms in a dwelling. The following were not counted as rooms: bathrooms, pantries, halls, clothes closets, and rooms used solely for business purposes. Summer kitchens, sunrooms, rooms in basements and attics were not counted unless they were occupied as living quarters throughout the year. 'Kitchenettes' were included if used for normal kitchen functions, and contained, as a minimum, cooking facilities (such as stove or range) and a sink or tub" (vol 3, xx). In 1961 we read: "Only rooms used or suitable for living purposes in the dwelling (including rooms occupied by servants, lodgers, or members of lodging families) are counted. Sun rooms, summer kitchens, recreation rooms, attic rooms, etc., are counted as rooms only if they are finished off and suitable for year round living quarters. Not counted as rooms are: bathrooms, clothes closets, pantries, halls, or rooms used solely for business purposes" (*1961 Census of Canada*, vol 2[2], xix). These varying definitions suggest some of the difficulties involved in identifying a room in practice. They also suggest caution in historical comparisons of data on room counts.

The floor area of the dwelling provides still another measure of size. However, because of practical difficulties in getting valid responses, few surveys inquire about this. Canada Mortgage and Housing Corporation (CMHC), in its 1974 Survey of Housing Units, equipped interviewers with tape measures to get room dimensions. Variations in how floor areas are to be measured make interviewer measurement almost obligatory for consistency. Variations occur in the practices of including or excluding exterior patios, balconies, garages, lanais, sunrooms, attached greenhouses, or lean-tos; unfinished or finished attic or basement areas; hallways, foyers, closets, and bathrooms; and floor areas covered by interior or exterior walls.

Measures of dwelling quality are discussed later in the book. Suffice it to say here that there are few survey data on housing quality. CMHC's Survey of Housing Units in 1974 did include some information about the state of repair of the dwelling (as perceived by the interviewer). The Census provides information on basic housing services such as running water and indoor toilets. In the 1981 Census, respondents were also asked to indicate whether, in their judgment, the dwelling required any minor or major repairs (excluding desirable remodelling or additions).

A HOUSEHOLD

In the 1941 Census of Canada, a household was defined to be "a person or group of persons living in one housekeeping community." Each (occupied) dwelling was seen to consist of one or more housekeeping units (or households). As stated in that Census (vol 5, xiii): "Two or more households may occupy the same dwelling. If they occupy separate portions of the dwelling and their housekeeping is entirely separate (including separate tables), they shall be numbered as separate households." The 1941 definition thus considers the social and spatial organization of homemaking activity within the dwelling.

In this book (as in the 1951 and subsequent censuses of Canada), a different definition is employed. A household is defined to be the collection of all individuals normally resident in a given dwelling.[4] These later censuses distinguish between households residing in private dwellings and those residing in collective dwellings. This book is concerned almost exclusively with private households and individuals. The adjective "private" is omitted for parsimony, except where households in collective dwellings are explicitly considered.

The 1941 Census definition gives more, and smaller, households than does the definition used in this book. Two households residing in the same dwelling, according to the 1941 Census definition, are here treated as one larger household. Thus it is difficult to compare counts of households from 1941 and earlier censuses with those of more recent years.

Census takers attach individuals to a dwelling where they are "usually resident" on census day (typically 1 June).[5] In later censuses, respondents were to indicate any "visitors" (i.e. persons not normally resident there) present on that date. Further, respondents were to include persons temporarily away but normally resident there. The purpose was to ensure that each person was counted exactly once.

Therein is the heart of a problem in using census data to look at household formation and housing demand. Every person must be allocated to a household and thereby a dwelling. If, for example, the person resides on a sailboat, the vessel is part of the housing stock in census terms. While this may be quite appropriate, it does not square with data on new housing construction that are concerned only with traditional forms of housing. Conventional statistics do not include the production of sailboats under new housing units. Hence, there is an inconsistency between housing stock change, as measured from the census, and housing stock change as calculated by summing new housing construction plus dwelling conversions plus net reduction in vacancies less demolitions.

Remember also that the census, and other surveys, are "snapshots." A snapshot survey asks about household and dwelling characteristics at the current moment. It does not ask if or how the composition of the household changed in the period leading up to the survey. Thus it can give a distorted picture of household formation and housing demand. For example, if an empty nester couple lives in a large dwelling, does it have a high level of consumption relative to its household size? Or does its dwelling choice reflect simply that a grown-up child had just left home or that another is about to move back? A snapshot survey misses elements of the dynamics of household formation that can be important in determining housing demand.

FAMILY

Throughout this book, "family" means a nuclear coresident family. By "nuclear" is meant a collection of only spouses/parents and never-married children. Married children are deemed to have formed their own families, as are never-married children who become lone parents. By "coresident," I mean that the family members live in the same household. Unmarried children, or spouses, who do not live with other family members are defined to be nonfamily persons. In other words, a family as defined here can be one of two following types: a husband-wife couple with or without never-married children at home, or a lone parent with at least one never-married child at home. Note that a child can be of any age. Also, a child need not always have lived at home. Some possibilities are illustrated in Figure 3. By definition, a family must contain at least two persons. This is the definition employed in postwar censuses in Canada; hence it is sometimes also referred to as a census family. In U.S. censuses, a different definition is employed. See the discussion below on "economic families" for more detail.

The census treatment of common law relationships has changed over time. In the 1976 and 1981 censuses, male-female couples living common law were included with husband-wife families.[6] Before 1976, censuses did not explicitly consider common law marriage.

According to the above definition, everyone must be either a family person or a nonfamily person. A family person is anyone residing in one of the two types of families described above. All others are nonfamily persons. Since a family must contain at least two persons, anyone living alone is automatically a nonfamily person. Anyone not in a husband-wife or parent-child relationship is also a nonfamily person according to this definition. Each family person must be a spouse, parent, or child to someone else in the same household.

These definitions of family and family status are arbitrary. The definition of a family is particularly narrow. It excludes, for instance, grandparents, uncles and aunts, cousins, nephews and nieces, and in-laws. I use these definitions here because many of the census, and other household survey, data are based on them. The narrowness of the definition does not, however, pose a substantial empirical problem. In 1981, two-thirds of all households in Canada consisted of a family living alone. Given that about one-fifth of all households are persons living alone, the small residual number of households are the only ones for which alternative definitions would make any difference.

Since 1956, the census has also employed the concept of an "economic family": i.e. a group of two or more coresident individuals related by blood, marriage, or adoption. Thus every census family living alone is also an economic family. An economic family can also consist of a census family plus other relatives such as grandparents, uncles and aunts, cousins, or in-laws. An economic family can also consist of any other combination of relatives, such as two brothers sharing a dwelling. An

FIGURE 3

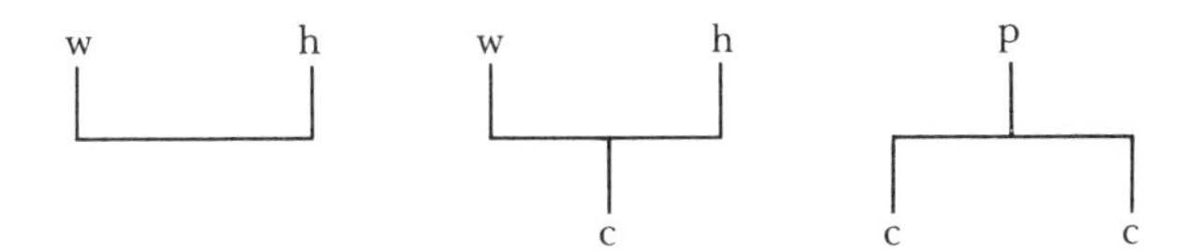

h: husband
w: wife
p: father or mother
c: never-married child living at home

economic family is a broader concept than is the census family. The U.S. definition of a "family" is equivalent to the Canadian "economic family": see Glick and Norton (1977, 26). In this book, occasional references are made to economic families. To avoid confusion, though, the term *family* always means a census family, unless prefixed by "economic."

This book focuses on patterns of household formation and housing demand. In part, it seeks to estimate empirically the role of demographic shifts on these patterns. To do this, however, we need an observational unit with two properties: rigidity and relevance. By rigidity, I mean that the measurement of the observational unit itself does not depend on the process being investigated. By relevance, I mean that the observational unit is insightful with respect to the process being studied. What is the observational unit in our analysis? Is it or is it not rigid? Is it relevant?

Two kinds of observational units are possible in such research. One unit is the individual. The number of individuals does not change as they are assigned to households; the individual is a rigid observational unit. The other observational unit is the set of families and nonfamily individuals. Because family members often make joint decisions about such things, the family may be a more insightful unit for the analysis of household formation and housing demand than is the individual. However, the family can be an elastic unit. Some persons, such as young adults, can and do choose between remaining in a family and moving out. In such circumstances, the family unit is itself a function of housing costs and housing market conditions. This is underlined by the emphasis on living arrangements (the term used is *coresident*) in the definition of a family. The danger in using families as the observational unit in a housing market study is that the number of families itself may be a response to housing market conditions. The family can thus be an elastic measure.

This is a fundamental problem in empirical research on household formation and housing demand. Individuals and families are the two obvious choices as units of measurement. However, it may not be insightful to look at the individual in isolation from the family. At the same time, the family is an elastic unit because its definition depends in part on living arrangement, itself the very process under study.

HEAD OF HOUSEHOLD

Because households take many forms, there is a need for some simple, effective ways of characterizing different cohorts wherein households have similar kinds of housing demands. A commonly used approach is to categorize households by the characteristics of its "head."

The term *head of household* is emotionally laden. Although survey takers use it in a statistical sense, the term has come to imply a power or authority structure within the household. The rise of the women's rights movement in the 1970s focused attention on the need to eradicate outdated sex stereotypes. An image of the husband as chief income earner, household provider, judge, and decision-maker was relentlessly attacked, and "head of household" was disputed as a manifestation of this image. In censuses up to 1971, the head of household was defined as: (i) the person living alone, (ii) the parent in a lone-parent family maintaining its own household, (iii) the husband in a husband-wife family maintaining its own household, or (iv) any person in a group of people sharing a dwelling as partners. This automatically excluded married women whose spouse was present from being head, regardless of her contribution to the maintenance of the household. A similar definition was employed in U.S. censuses through 1970. In the 1976 Census of Canada, condition (iii) was modified to be either husband or wife. A similar concept, the "householder," was employed in the 1980 U.S. Census.

In the 1981 Census of Canada, head was dropped entirely. In the questionnaire, "Person 1" was used instead. "Person 1" was defined as either spouse in any married couple present in the household, either partner in a common law relationship, the parent where one parent only lives with his/her never-married children, or, if none of the above is applicable, any adult member of the household. The 1981 questionnaire also asked for the name of the person who "is responsible for paying the rent, or mortgages, or taxes, or electricity, etc., for this dwelling." That person is referred to as the "household maintainer." This person need not be resident in the household: e.g. a separated husband who is supporting a wife and children. In some published 1981 Census reports, tabulations are presented by the characteristics of the household maintainer: e.g. age. In such cases, the household maintainer is defined somewhat differently; it is the household maintainer as above if coresident, or Person 1 otherwise.

Why put so much attention on the characteristics of the head of household? A common argument is that most individuals share a common life cycle. They are born, grow up, leave home, get married, have children, launch these children when they become adults, get older, and die. Many people tend to experience various life cycle events (e.g. marriage, birth of first child, marriage of last child, death of spouse, own death) at about the same ages. Therefore the age of the head or spouse is a crude predictor of his or her life cycle position. For example, a married (spouse present) male head aged 42 in 1982 might be expected to have wife aged 39, have been married about 17 years, and have an eldest child about 13 and a youngest about 10. Of course, there are always exceptions: the person who married late, who remained childless, or whose child spacing pattern was different.

After about 1960, new trends raised questions about the relevance of the family life cycle. The decline in the propensity to marry among young single adults, the decline in fertility and rise in childlessness, and the increasing incidence of divorce all imply that the life cycle changed in some way for many individuals. In some cases, the temporal ordering of life cycle events was unchanged but the ages at which given events occurred did change. In other cases, however, the rearrangement or disappearance of certain life cycle events became more frequent. Although many people continued to be typified by the traditional life cycle, an increasing number did not.

HOUSEHOLD TYPE

The age, sex, marital status, and family status of the household head are not the only important indicators of housing needs or preferences. In housing demand studies, other common indicators focus on household size, presence of children at various ages, expectations about near-future child birth, and family composition.

One focus of this book is on the decline in shared accommodation. Therefore it is useful to have a household type disaggregation that allows measurement of this. We want to be able to count the number of families living alone versus the numbers in different forms of shared living arrangement. Also, in households with a mix of families or of family and nonfamily persons, it is helpful to know who is maintaining the dwelling and who is lodging.

Figure 4 shows the disaggregation of households possible in these terms. Using the pre-1981 census concept of a household head, seven household types emerge, labelled (a) through (g). 1981 Census data are based on household maintainer. For many households, "head" and "maintainer" are the same person. There is a new wrinkle, though, in that a maintainer does not have to reside in the household. For example,

FIGURE 4

Families present	Nonfamily person(s) present	Household head: Family member	Household head: Not family member
None	Yes	n.a.*	(a)
One	No	(b)	(1)
	Yes	(c)	(d)
Two or more	No	(e)	(2)
	Yes	(f)	(g)

* Not applicable

a separated woman and her children living alone, but supported by a husband, would have been treated as a category (b) household before, but as a category (1) household in 1981. Thus the household maintainer concept raises the possibility of two additional household types, labelled (1) and (2). These two types are not possible under pre-1981 household head definitions.

For various reasons, including parsimony and data availability, this book will often consider fewer than these seven or nine types. Typically, just three types are used: primary nuclear family living alone (PNF), primary nuclear family with others present (PNFP), and households without a primary nuclear family (NPNF). A PNF household is of type (b). About two-thirds of all households in Canada in 1981 were of this type: i.e. families living alone. A PNFP household is one of types (c), (e), or (f); the head of household is a family member, but there are other nonfamily persons or families lodging there as well. An NPNF household is one of types (a), (d), or (g), wherein the household head is not a family member. In the 1981 Census, this type also includes nonresident household maintainers as in types (1) and (2). Persons living alone are included in (a) and account for most of the postwar NPNF households.

This categorization helps to clarify patterns of change in household formation and housing demand. The categories distinguish between families living alone and other kinds of households. They distinguish also between households where there is a primary family and those where the head is a nonfamily person. The explanations presented in chapter one emphasize the decline of doubled-up families and the rise of the nonfamily individual. The categorizations used here allow a better separation of the roles of these explanations.

CONCLUDING COMMENTS

Several themes are expounded in this chapter. First, there is the importance of the concept of "privacy." The definitions given for a dwelling and a room are predicated on a concept of privacy, whether it be in terms of unshared facilities, separating walls, or private entrance. In later chapters, we will return to this again, because it has been argued that part of the changing pattern of household formation and housing demand can be attributed to preferences for "privacy." What is privacy? Why is it important? How has it shaped household formation and the demand for housing?

Second, the difficulty of deriving empirical measures of housing quality and size is discussed. Although we can often quickly develop a subjective assessment of the size and quality of a given dwelling from a myriad of detail, it is another matter to come up with a few objective measurements that do the same thing. However, a decomposition of the shelter cost of a dwelling into price and quantity-quality components is essential to meaningful analysis of housing demand.

Third, census data focus on the principal residence occupied by a household. For the most part, secondary residences are ignored. However, an analysis of housing demand should include some consideration of the extent of substitution among alternative housing forms, including secondary residences.

Fourth and finally, the chapter begins a discussion of alternative ways of characterizing households that might have similar patterns of housing demand. One way is to look at the characteristics of the household head. There are strong connections between life cycle changes – often associated with changes in housing preferences – and age. However, changes in its definition over time make intertemporal comparisons based on household headship somewhat unreliable. This chapter has explored an alternative route that categorizes households by family composition. Its utility is illustrated in later chapters.

CHAPTER THREE

Postwar Patterns of Family Formation

The early postwar years witnessed trends toward younger and more prevalent marriage, earlier first child births, and larger completed family sizes. Later, the trends were to be reversed: later marriage, more bachelorhood, more divorces, fewer children, and postponed child birth. As a result, family formation slowed, and the demographic characteristics of families changed. New living arrangements emerged, along with new housing forms to accommodate them.

This chapter explores postwar family formation. The purpose is to develop some impressions of how and why family formation changed. Different explanations are considered, and the evidence is reviewed. Chapter four suggests that changes in household formation, composition, and size can be explained partly in terms of changes in family formation. An understanding of family formation is therefore important in comprehending how people chose their living arrangements.

The description begins with a discussion of postwar population change. The effects of changes in fertility, mortality, and migration are noted. Next, we examine changes in nuptiality, since marriage typically precedes family formation. This includes an assessment of the effect of changes in nuptiality on the incidence of bachelors and ever-marrieds and a discussion of widowhood, separation, divorce, and remarriage. Finally, we assess the effects of these population and marital distribution changes on family formation in terms of types and sizes of family units.

POPULATION CHANGE

The pace of population growth in postwar Canada was especially quick up to about 1961. From 1941 to 1961, population increased at an average of 2.3 per cent annually. The pace of growth slowed after 1961, rising at just 1.4 per cent annually through 1981. We are interested in the impli-

TABLE 2
Population (000s) by family status living arrangements, Canada, 1941-81

	1941	*1951*	*1961*	*1971*	*1981*
In private dwellings					
Family members					
Living with spouse*	4,432	5,923	7,600	9,184	11,222
Lone parent	309	326	347	479	714
Child					
24 or younger	4,693	5,544	7,778	8,855	8,253
Over 24 years old	451	423	371	334	414
Nonfamily individuals†	1,237	1,384	1,659	2,323	3,195
In collective dwellings	368	384	484	393	406
Total (usual residents)‡§	11,490	13,984	18,238	21,568	24,203

SOURCE: Censuses of Canada, various years

* In 1981, common law couples were enumerated as marrieds. In earlier censuses, when such couples did not list themselves as married they were counted as either nonfamily individuals (if no children present) or lone-parent families (if children present). Thus, the 1981 Census estimates more husband-wife families and fewer lone parents and nonfamily individuals than previously.

† Includes individuals whose family status could not be ascertained.

‡ Includes small number of individuals living within families in collective dwellings.

§ Columns may not total due to rounding.

cations of this population growth for household formation and housing demand. Consequently, it is valuable to distinguish between total population and the population of individuals in private dwellings (as described in chapter two). Data presented in Table 2 are telling. Collective dwellings include work camps, barracks, staff or student residences, hospitals, and hotels. The number of individuals in collective dwellings remained stagnant between 1941 and 1981. Thus, in addition to overall population growth, a rising proportion of Canada's population was being accommodated in private households. The significance of this shift is discussed further in chapter four. Also, the table illustrates that the vast majority of Canadians lived in families, although nonfamily individuals became relatively more common over time.

To better understand what happened in terms of aggregate growth, Canada's population growth can be decomposed into three components: births, deaths, and migration. Each of these components changed in distinctive ways, and population growth can be better understood by examining them separately.

Fertility

Modern vital statistics record keeping began in Canada in the 1920s. From then until 1939, the number of births varied only slightly, from

230,000 to 250,000 per year. After the start of the Second World War, fertility rose steadily: reaching a plateau in the late 1950s at just under 480,000 births annually. Because of this, the period roughly from 1946 to 1960 is often called the postwar "baby boom." By 1963, however, the birth rate had fallen to 364,000 births. It continued to fall and then stabilized in the 1970s at about 340,000 to 360,000 annually. The period since 1960 has been dubbed the "baby bust." Although the baby bust continued at the time of writing, the term is used here to refer to the 1960s and 1970s.

Some other developed countries also experienced a postwar baby boom of the magnitude and duration of Canada's: namely the United States, Australia, and New Zealand. However, in other developed countries – notably in Europe – there was no such fertility boom. Whatever social processes generated the fertility boom, they were not universal.

These data on births give an impression of the magnitude of postwar baby boom relative to the birth cohorts before 1940 or in the baby bust period. At its peak, there were approximately twice the number of births experienced per year as in the prewar period, and 30 to 40 per cent more than in the 1970s.

Data on the total number of births do not control for the changing composition of Canada's population. In 1951, there were about 1.7 million women of prime child bearing age (20–34). As late as 1966, there were still under 1.9 million. Then, as the baby boom cohort aged, these ranks swelled: rising to 3.3 million by 1981, in sharp contrast to the modest growth in total population at this time. If the propensity of these women to have children had remained constant, the number of live births should have increased accordingly. That it did not makes the decline in total births after 1960 even more dramatic.

Rather than use total births, demographers prefer measures based on fertility rates. A fertility rate is the average number of children born in a year to 1,000 women of a certain child bearing age. Fertility rates for Canadian women are shown in Table 3. As an example, there were 100.1 births in 1980 per thousand Canadian women aged 20 to 24.

Over time, fertility changed, although the pattern varied considerably by age group. Women aged 40 or older had steadily declining fertility rates after 1921, even through the baby boom period. The fertility rates of 35–39-year-old women also did not increase during the boom. It is only among the under-35s that fertility rates peaked. In this age group, fertility rates had declined from 1921 to 1940, then rose through to about 1960, and subsequently declined.

The period total fertility rate (PTFR) is the number of births expected per thousand women who pass through their lives with a given set of age-specific fertility rates. Consider the 1940 fertility rate data above as an

TABLE 3
Age-specific fertility rates for Canadian women, 1921-80

	Age group of women						
Year	*15-19*	*20-24*	*25-29*	*30-34*	*35-39*	*40-44*	*45-49*
1980	27.6	100.1	129.4	69.3	19.4	3.1	0.2
1978	29.7	103.1	128.1	67.1	19.5	3.6	0.3
1960	59.8	233.5	224.4	146.2	84.2	28.5	2.4
1940	29.3	130.3	152.6	122.8	81.7	32.7	3.7
1921	38.0	165.4	186.0	154.6	110.0	46.7	6.6

SOURCE: Published vital statistics reports, various years

example. If 1,000 women were to pass through the ages of 15 to 49 giving birth at the annual rates shown above, they would experience a PTFR of 2,766 births.[1] The PTFR for Canadian women declined from from 3,536 births in 1921 to 2,654 in 1939, then increased to a peak of 3,925 in 1957, and subsequently declined to 1,746 in 1980. The Population Reference Bureau (1982, 7) reports a similar pattern in the United States: a PTFR of about 3,600 in 1900, falling to 2,100 in 1936, rising to 3,200 in 1947 and 3,700 in 1957, and then falling after 1961 to just under 1,800 by 1976.

A second baby boom (sometimes referred to as a "mini boom," "echo boom," or "boomlet") was widely expected in the 1970s and 1980s as the original baby boom cohort moved into its child bearing years. That boom had not materialized in Canada by 1981. Fertility rates declined about as fast as the number of women of child bearing age increased. At best, the baby boom cohort was, by its sheer numbers, able to keep the number of births from falling. In the United States, the Population Reference Bureau (1982, 7) reports a slight upswing in fertility in the late 1970s - enough to raise the annual number of births from 3.1 million in the mid-1970s to 3.6 million in 1980.

Underlying the fertility shifts were changes in the timing and spacing of births. One important change was the decline in child birth among older women. Other changes become apparent when fertility rates are disaggregated by birth order. Of 100 live births per thousand women aged 20-24 in 1980, 58 were first births, 33 were second births, 8 were thirds, and 2 were fourth births or higher. A breakdown of fertility rates is given in Table 4 for various years from 1941 to 1980.

In panel (a), the postwar baby boom shows up partly as a decline in the typical age at birth of first child. Note the decline in first birth fertility rates for women aged 25-34 and the increases for women aged 15-24 during the 1950s. Note also the reversal during the baby bust, as women began in greater numbers to have their first child in their late twenties or early thirties. As well, note the decline in first birth fertility rates among

TABLE 4
Fertility rates, birth order* by age of mother, Canada, 1941-80

Year	*Age of mother* 15–19	20–24	25–29	30–34	35–39	40–44
(a) First births						
1980	23.0	57.6	51.5	16.4	3.4	0.5
1976	27.3	60.1	46.0	13.6	3.2	0.5
1971	33.2	71.0	43.0	11.4	3.0	0.6
1966	38.4	81.8	35.5	10.6	3.8	0.9
1961	42.7	94.1	39.3	13.6	4.8	1.2
1956	42.1	95.2	46.4	16.2	6.3	1.6
1951	37.1	86.2	45.5	17.8	6.9	1.7
1941	24.6	68.3	51.9	24.4	6.6	1.3
(b) Second births						
1980	3.9	32.7	51.1	26.5	5.8	0.6
1976	5.5	38.4	53.4	23.8	5.2	0.7
1971	6.2	45.2	50.4	19.0	5.1	1.0
1966	8.4	54.8	49.0	19.1	6.6	1.5
1961	12.7	76.5	56.6	23.7	8.1	1.9
1956	11.5	71.9	61.2	27.5	10.4	2.3
1951	9.2	60.1	59.6	31.1	12.1	2.5
1941	5.1	35.3	40.5	25.2	8.3	1.5
(c) Third births						
1980	0.4	8.1	20.3	16.7	5.0	0.6
1976	0.5	9.5	22.1	16.7	4.8	0.6
1971	0.7	13.6	28.8	19.3	6.2	1.1
1966	1.2	21.7	37.5	23.1	9.7	2.0
1961	2.4	39.3	52.0	30.5	12.5	2.9
1956	1.9	34.8	49.4	32.3	14.3	3.3
1951	1.6	27.0	42.4	30.4	14.4	3.3
1941	0.9	16.1	26.4	20.3	8.6	1.8
(d) Fourth births or higher						
1980	0.0	1.7	6.5	9.7	5.2	1.5
1976	0.0	2.2	8.3	11.5	7.9	2.5
1971	0.1	4.7	19.7	27.4	19.3	6.7
1966	0.2	10.9	41.5	50.5	37.4	14.7
1961	0.4	23.7	71.3	77.1	55.7	22.5
1956	0.3	20.4	63.1	74.3	58.7	23.7
1951	0.2	15.4	51.3	65.2	53.1	23.4
1941	0.1	18.7	41.0	52.5	56.5	27.1

SOURCE: Computed from vital statistics (various years) using age-specific rates and births by age of mother and birth order.

* Only live births are included here. Data on birth order for 1941 available only for legitimate births. In this table, these 1941 data were "blown up" by the ratio of total to legitimate births for 1951 for the corresponding age of mother and birth order.

older women. Thus the baby bust period was characterized by an increased concentration of first births in a narrow range of ages of mother.

During the baby boom period, women also tended to have their second child at a younger age than before. In panel (b), this is especially evident among women under 30. These rates fell again after 1961. Note, after about 1971, the rising incidence of women having their second child only when they are 30-34. The baby bust period was characterized by a delay in having both the first and the second child.

During the baby boom period, it became more common for women to experience a third child birth. In the baby bust, third births dropped back to 1941 levels (or even lower). In large part, much of the baby boom was attributable to decisions to have three instead of two children. Additional support is evidenced in panel (d), which shows that higher-order fertility (fourth birth or higher) did not increase much during the baby boom. After about 1961, however, these higher-order fertility rates fell quickly. The baby bust period can be characterized by a return to 1941 third birth fertility rates and a drastic reduction in higher-order fertility.

The above data give an impression of postwar fertility changes among women who had children. They do not, however, show the incidence of childlessness. Such data are unavailable for Canada. However, Heuser (1976) estimates the probabilities that U.S. women by five-year birth cohorts will be of a specified parity (i.e. have given birth to a specified number of children) by a particular age. Heuser found that the percentage of women reaching age 50-54 without a live birth hovered around 18 to 20 per cent among white women born between 1886 and 1912. However, among the cohorts of women born later, the incidence of childlessness fell dramatically. For example, among women born between 1920 and 1924 - one of the cohorts principally responsible for the postwar baby boom - fewer than 10 per cent were childless. Bloom (1982) presents data, again for the United States, that show the proportion of childless women will likely rise for women born after 1935. These are the women responsible for the baby bust. Bloom projects that perhaps 30 per cent of women born in 1955 will not bear children. Two conclusions follow from this. First, a sharply reduced incidence of childlessness helped to create the postwar baby boom. Second, an even sharper rise in childlessness (potentially to levels unprecedented in this century) subsequently helped to create the baby bust.

It is facile to think of the baby boom as an aberration and the subsequent baby bust as a return to "normality." It is correct in that, in the baby bust period, fertility continued a downward trend evident in the first four decades of this century, that there was a return to prewar patterns of age of mother at child birth and away from the youthful child bearing of the baby boom period, and that the decline in fertility among older women continued. However, the baby bust period was more than just an unravelling of the baby boom. Fundamental and unprecedented changes in fertility included a rising, possibly unprecedented, incidence of childlessness and a sharp reduction in fourth and higher-order births compared to both prewar and baby boom years.

These postwar fertility swings remain one of the great puzzles of modern demography. A variety of explanations have been offered for them. Let us briefly recount some of these.

Several explanations assert an economic rationale. Fertility behaviour is seen to result from a conscious choice in which parents weigh the costs and benefits of having children. Sometimes, these explanations concern the best starting age for parenthood and the best completed family size. Other explanations focus on the incremental decisions of parents to have an additional child. Whatever the focus, these explanations see fertility behaviour as a conscious, rather than random, choice.

Some of these economic explanations are linked to the seminal work of Becker (1960). Becker emphasizes how a family uses its available time and other resources to produce "outputs" including, but not limited to, children. Becker's approach emphasizes that parents must devote time, and other resources, to child rearing: resources that could be used for other purposes in the absence of children. Empirical applications of this approach, based on U.S. data, include Lindert (1978) and Turchi (1975). Such studies emphasize the importance of opportunity costs such as the impact of child rearing on the labour force participation of both parents, forgone employment income and its relationship to each parent's education, the marginal out-of-pocket costs of raising an additional child, and tax treatment of dependants.

A different economic explanation has been offered by Easterlin in several articles culminating in a little book in 1980 entitled simply *Birth and Fortune*. His "relative economic status hypothesis" has three parts. First, under prevailing social norms, the most desirable family sizes are between two and four children. Within this range, the couple is usually indifferent. Second, the number of children a couple will have (between two and four) is determined by the number of fertile years covered by the marriage. The older a woman at the time of marriage, the fewer children (i.e. the closer to two) can be expected. Third, in deciding when to marry, young couples weigh their economic status relative to that of their parents at the time they left their parents' homes. The lower their relative economic status, the more likely they are to delay marriage.

Easterlin presents data to show that during the baby boom period, the incomes of young adult males increased quickly relative to the incomes of their fathers. This, he argues, induced earlier marriages and hence greater fertility. During the baby bust period, the incomes of young adult males fell relative to those of their parents, thereby inducing fewer marriages and lower fertility. Easterlin argues that the incomes of young adults have varied with respect to their elders because of relative supply. Because of low prewar fertility there were relatively few young males entering the labour force in the late 1940s and the 1950s to compete for available jobs. However, the baby boom cohort, because of its swollen numbers, experienced more unemployment and sluggish income growth.

Several noneconomic rationales have also been put forward to explain fertility changes in the postwar period. Some adopt the part of Easterlin's argument that ties fertility to age at marriage. Unlike Easterlin, they do not argue that the decision to marry earlier is economically determined. They suggest that other factors led couples to marry at younger ages during the late 1940s and the 1950s. Whatever the cause, higher fertility rates are seen as a consequence of this rising propensity for early marriage. Similarly, the authors see the baby bust arising because of delays in marriage.

Still others say that cultural values and preferences changed in the postwar period. It is argued that couples in the baby boom period wanted, and therefore had, more children. In the baby bust period, though, conditions changed. A heightened awareness of alternative life-styles and increasing public concerns over environmental preservation and overpopulation led to smaller preferred family sizes. Some support for this can be found in national polls of desired family size. Blake (1974, 27) reviews data taken from Gallup opinion polls taken since 1945. The Gallup poll asked the following question; "What do you consider is the ideal size of a family - a husband, wife, and how many children?" Among young white U.S. adult women, the average response fluctuated between 3.0 and 3.5 children in surveys taken from 1945 to 1967. In the early 1970s, the average dropped to just 2.7 to 2.9. These data indicate a changing concept of desired fertility. However, it is unclear how and why these preferences changed.

Finally, the acceptance of new contraceptive methods, particularly the oral contraceptives that appeared around 1960, is argued to have been an important cause of the baby bust. Such contraceptives helped couples to bring their actual fertility down to a desired level. However, it has been noted by critics that fertility in the 1930s had been very low without such contraceptive devices. In their view, the sudden reduction in fertility after 1960 cannot be explained solely in these terms.

Evidence of the role of contraceptives is offered in Ryder and Westoff (1972). They used U.S. data from national fertility studies to measure the extent of unwanted births. Women interviewed in these surveys were asked several questions about their desired completed family size and about the desirability of each birth as it occurred. They found that women married in the period 1941–5 had, on average, 0.6 unwanted births. The average number of unwanted births fell sharply for women marrying after about 1956. For women marrying between 1961 and 1965, for example, the average was just 0.1 unwanted births. Given that the decline in cohort total fertility during the 1960s was about 1 birth per woman, the decline in unwanted births of about 0.5 births per married woman is significant. Thus Ryder and Westoff conclude that improvements in contraceptive technology were important contributors to declining fertility during the baby bust.

One should also consider the role of abortion. Canada's abortion laws were liberalized in the late 1960s, making abortions legal under certain circumstances. Data on the impact of abortion on Canadian birth statistics are unavailable. Glick and Norton (1977, 24) report on the U.S. experience between 1970 and 1975. They find that induced abortions (legal and illegal) accounted for under 13 per cent of all pregnancies in 1970. By 1975, this had risen to 20 per cent, almost all of which were legal abortions. They estimate that the total number of pregnancies in the United States rose between 1973 and 1975. However, because of the rapid increase in legal abortions, the number of live births actually declined. Canada's abortion legislation is somewhat more restrictive; hence the impact of liberalized abortion was likely somewhat smaller.

Mortality

In Canada between 1945 and 1976, life expectancy at birth increased by 5.5 years for men (from 64.7 to 70.2 years) and 9.5 years for women (from 68.0 to 77.5 years). This increasing longevity itself was a source of population growth. The longer people live, the more likely they are to complete their child bearing years and still to be alive when their grandchildren or subsequent generations are born.

What effect did increasing longevity have on Canada's total population? This is difficult to answer exactly because of the time required for the effect of changed mortality rates to be fully felt. For example, a decline in infant mortality today will eventually affect the number of persons aged 70, but we have to wait 70 years to witness it.

To address such questions, demographers use a life table. This shows the effect of a regime of mortality rates (one rate per year of age) on a "steady state" population: i.e. wherein there is neither immigration nor emigration and where fertility rates are just high enough to hold the total population constant. Also, a life table represents a "steady state" in that it is calculated under the assumption that the population at any age has already experienced the current mortality rates of all younger ages.

TABLE 5
Life expectancy data for Canada, 1930-2 to 1975-7

	Male				*Female*			
Life table	*0s1**	*e1†*	*1s60‡*	*e60§*	*0s1*	*e1*	*1s60*	*e60*
1930-2	0.913	64.7	0.722	16.3	0.931	65.7	0.735	17.2
1940-2	0.938	66.1	0.747	16.1	0.951	68.7	0.787	17.6
1945	0.943	67.6	0.770	16.4	0.954	70.3	0.812	18.1
1950-2	0.957	68.3	0.778	16.5	0.966	72.3	0.847	18.6
1955-7	0.965	69.0	0.794	16.5	0.972	74.0	0.872	19.3
1960-2	0.969	69.5	0.800	16.7	0.976	75.0	0.885	19.9
1965-7	0.975	69.5	0.798	16.8	0.980	75.7	0.889	20.6
1970-2	0.980	69.8	0.803	17.0	0.985	76.6	0.893	21.4
1975-7	0.985	70.2	0.810	17.2	0.988	77.4	0.901	22.0

SOURCE: Computed from life tables for Canada in various years, as published by Statistics Canada
* Probability of live birth surviving to first birthday
† Remaining life expectancy at first birthday, in years
‡ Probability of surviving from first to sixtieth birthday
§ Remaining life expectancy at sixtieth birthday, in years

Thus, a life table shows the long-run, or full, effects of a particular mortality regime. Life tables are available for Canada for several years. The ratio of the steady state population for Canada in the 1976 life table, compared to the 1945 life table, is 1.11.[2] This means that the regime of mortality rates in 1976, in a steady state, would yield a population 11 per cent larger than that given by the 1945 regime. This is a rough, but useful, estimate of the effect of increasing longevity. Remembering that Canada's population almost doubled over this time period, the contribution of increasing longevity was not insubstantial.

Improved longevity came about principally via two sources: reductions in infant mortality and reductions in mortality among the middle aged and elderly. The reduction in infant mortality was especially marked. The 1931 life table shows that infant mortality (the probability of death before one's first birthday) was 8.7 per cent for males, compared with 1.5 per cent in 1976. The probability of a one-year-old male dying before his sixtieth birthday fell somewhat less rapidly, from 28 per cent in 1931 to 19 per cent in 1976.

The two sexes did not share this increasing longevity equally. As illustrated in Table 5, men traditionally had higher mortality rates than women of the same age. This differential shows up, for example, even in infant mortality rates, which were about one quarter again as large for males as for females. During the postwar period, the gap widened. In 1931, a one-year-old female could expect to outlive a corresponding male by one year. By about 1976, this had risen to seven years. This relative improvement in female longevity permeated all age groups; it was not exclusive to the elderly. Women were increasingly more likely than men to reach an older age and to outlive men after that. It is difficult to overemphasize the effect of this increasing sex differential in longevity on household formation and housing demand. With increasing longevity, relatively more elderly persons require accommodation, whether in their own dwelling or a shared unit.

Immigration and Emigration

Since the beginning of European colonization in the seventeenth century, Canada has experienced waves of immigration. In the first decade of this century, for example, immigration accounted for almost one-half of total population growth, then subsided during the next four decades. In the postwar period, large volumes of immigration occurred during the 1950s and again from 1965 to about 1974. These waves make it difficult to describe a "typical" level of immigration. In 1980, for example, there were about 129,000 immigrants, but since 1960 the annual volume has varied from 70,000 to 214,000. Because of this ebb and flow, the importance of immigration in Canada's population growth has varied considerably. However, in the 1970s, immigration became more important as fertility declined. In 1974, for example, immigration was almost 60 per cent of the number of live births.

Relatively little is known about emigration from Canada, since there is no formal system of registration for emigrants. Statistics Canada has, since 1961, estimated emigration using the immigration records of recipient countries. The estimate for 1980 is 74,000 emigrants, and estimates have ranged from 62,000 to 111,000 for the years since 1961. These estimates show a wave pattern, similar to that for immigration, but more muted. Net immigration, immigration less emigration, can thus be estimated only for the years since 1961. In the early 1960s, net immigration was negative (more emigrants than immigrants). However, it became positive around 1963 and continued to increase until the start of the recession in 1968. From a peak of over 110,000 persons per year in 1966-7, it fell gradually to about 51,000 by 1971-2. Another peak in 1974-5 at about 143,000 was followed by a decline to 55,000 net immigrants in 1980. Net immigration was an important source of overall population growth in Canada.

PATTERNS OF FIRST MARRIAGE

By "first" marriage is meant the marriage of single individuals. First marriage differed in the early and later postwar years. The early period has

TABLE 6
Proportion single of 25-34-year-olds, Canada, 1891-1981

Year	*Men*	*Women*
1891	0.436	0.298
1911	0.463	0.269
1921	0.378	0.232
1931	0.413	0.259
1941	0.400	0.275
1951	0.276	0.174
1961	0.233	0.129
1966	0.212	0.121
1971	0.200	0.125
1976	0.207	0.131
1981	0.237	0.154

SOURCE: Computed from the census of Canada, various years

been called the "marriage rush." I refer to the later period as the "marriage bust." There is no exact dividing date. The transition from marriage rush to bust took place slowly throughout the 1960s. Compared to earlier in the twentieth century, people in the marriage rush were more likely to marry, marry at a young age, and marry close to a modal age. In the subsequent marriage bust period, marriage patterns began to shift slowly back towards those of the prewar years.

Mean Age at First Marriage and Bachelorhood among Young Adults

One indicator of these changes is median age at first marriage for a specified birth cohort. Rodgers and Witney (1981, 729) estimate that for cohorts born in the 1880s through the 1920s, the median age hovered around 23 years for women and 26 years for men. However, during the marriage rush, these medians dropped to about 21 and 24 years respectively (for the cohort born in the 1940s). During marriage bust, the median slowly began to turn up again. For women born in the 1950s, it rose to 21.5 years.

The popularity of youthful marriage up to 1970 is evident in Table 6. Among men, between 37 per cent and 46 per cent of 25-34-year-olds were enumerated as never married in pre-Second World War censuses. However, after the war, this dropped to 28 per cent in 1951, then to 20 per cent in 1971. Among women, the probability of a 25-34-year-old still being single in 1971 was only one-half what it had been in 1941. Modell et al (1978, s123) report similar findings in the United States.

The late 1960s and early 1970s marked the start of a reversal. The proportions of young adult men and women remaining single increased in 1976 and again in 1981. The trend away from marriage (or, at least, early marriage) was small compared to the incidence of bachelorhood in the prewar decades. Nonetheless, it was the first decline in youthful marriage in the entire postwar period. A similar shift in the United States is reported in Bernard (1975, 584).

Marriage Rush and Marriage Bust

Modell et al (1978) argue that, in the marriage rush in the United States, there was also a reduced variation in age at marriage. They call this reduction the "increasing modality" of marriage. In 1928, for example, the spread between the 10th and 90th percentiles of age at first marriage was 14 years. By 1969, the spread had dropped to just 10 years. Modell et al (1978, s128) comment: "The postwar pattern of more modal marriage, initially a function largely of generally heightened probabilities of marriage, was now reinforced by a new age-specific marriage schedule. Older marriage probabilities became somewhat depressed absolutely even as younger singles increased their marriage probabilities. Viewed from a life-course perspective, marriage is always something of a segregation process: those who by a particular age fail to marry find their subsequent chances of ever marrying narrowing each year. Between the postwar rush and the end of the 1950s, this selective process became even more pronounced."

There is evidence of a "bunching up" of marriages by age group in Canada during the marriage rush. In Table 7 are presented first marriage rates (FMRs); these are the number of first marriages in a year in ratio to the number of singles enumerated in the census on 1 June of that year. For example, the number of single men aged 30–34 who married in 1956 was 10 per cent of the number of single males enumerated in the Census of 1 June.[3] Modell's argument suggests that we should begin with pre-Depression data, but these are unavailable. We begin instead with 1941, the earliest date for which data are available. The marriage rush period shows up in this table as a bunching of high FMRs in the younger age groups. Among males aged 15–19, the high postwar FMRs lasted well into the 1970s. For males 20–29, the FMRs peaked around 1966 and fell sharply during the 1970s. Similar results show up for women under 25. Modell's "increasing modality" shows up as a simultaneous reduction in FMRs among older age groups up to about 1961.

These FMR data clarify the source of the "marriage bust." The decline in propensity to marry, after about 1971, was most marked among younger (i.e. under 30) men and women. For all cases except men aged

TABLE 7
First marriage rates* by age and sex, Canada, 1941-81

	Men					*Women*				
Year	*15-19*	*20-24*	*25-29*	*30-34*	*35-39*	*15-19*	*20-24*	*25-29*	*30-34*	*35-39*
1941	0.005	0.095	0.068	0.139	0.086	0.045	0.166	0.168	0.102	0.058
1951	0.013	0.136	0.174	0.123	0.069	0.066	0.204	0.161	0.087	0.046
1956	0.013	0.144	0.173	0.100	0.056	0.071	0.222	0.165	0.082	0.042
1961	0.012	0.148	0.168	0.092	0.048	0.063	0.223	0.144	0.071	0.036
1966	0.012	0.158	0.196	0.098	0.049	0.057	0.222	0.154	0.068	0.036
1971	0.013	0.157	0.185	0.103	0.054	0.054	0.220	0.147	0.074	0.039
1976	0.011	0.118	0.150	0.093	0.052	0.042	0.168	0.135	0.071	0.040
1981	0.006	0.090	0.138	0.090	0.048	0.027	0.144	0.133	0.069	0.033

SOURCE: Data on marriages taken from vital statistics (various years). Data on population by age, sex, and marital status taken from the census of Canada (various years)
* Ratios of marriages of single persons during year to the number of singles enumerated in the census at mid-year

25-29, the FMR in 1981 was at or below the 1941 level. The effects of the marriage rush on the FMRs had been effectively erased by 1981.

Why did the propensity to marry change? Was the marriage bust simply a manifestation of rising cohabitation, as some people suggest, or are there alternative explanations? What are the relationships between the marriage bust and the baby bust?

What Caused the Marriage Swings?

Modell et al (1978, s143-s147) speculate on the origins of the marriage rush. They argue that the marriage rush originated in the Depression, not in the war years that immediately preceded the rush: "Beginning with the depression and continuing for a decade or more, both marriage and child bearing generally became problematic. Perhaps half a generation of youth thus were cut loose from traditional timing criteria, long enough for the criteria themselves to shift." In their view, the Depression created the possibility of massive changes in the timing of marriage and childbirth, and the Second World War merely delayed the start of this new life cycle pattern.

What determined the exact shape of the new marriage rush? Modell et al see three important factors. First, there was change in the typical family economy. Marriage in earlier times was often predicated on the ability of the husband to support a family as sole breadwinner. However, the rapid expansion of jobs during the postwar period meant that increasingly both spouses worked. The income necessary to sustain a marriage (and the separate dwelling that it increasingly implied) could be achieved typically at a younger age. Also, Modell et al argue that with the general prosperity of the postwar period, parents were able to support their married children financially: "A charming piece of historical ethnography, we suspect, would be to recreate the development (in the postwar period, probably) of the variety of ritual falsehoods by which parents subsidized the marriages of their children, who would have been both too young and too dependent to marry by the standards of an earlier generation."

Second, Modell et al argue that a variety of social assistance programs such as unemployment insurance made it easier for a family with a low income to support themselves. To their list, I would add family allowances, student grants, public housing programs, public medical insurance, and old age pensions, all of which have served to supplement incomes or reduce expenses for low-income individuals and families. Third, they argue that the "increasing modality" of marriage was attributable to a greater "age grading" in the postwar period. Because of the more uniform tendency of individuals to complete high school and college, they argue that individuals became more likely to enter the job market, and subsequently marry, at narrowly bunched ages.

These arguments are interesting. However, while consistent with trends in the marriage rush period, they do not satisfactorily explain the marriage bust. During the 1970s, the incidence of both spouses working continued to rise. I suspect that parental financial support of young couples also continued, although data are scarce. Also, social assistance programs continued apace. Finally, patterns of school leaving continued in the 1970s much as in the 1960s. What, then, brought about the marriage bust of the 1970s?

Another common explanation is that the marriage bust was the result of increasing common law marriage. Unfortunately, there are few data available on this. The 1981 Census was the first in which common law marriage was a distinctly identified category. Prior to 1981, a common law couple might have listed themselves as "married" or might instead have chosen some other category such as "partners." There is simply no way of identifying common law arrangements exactly from these earlier census returns. This is further complicated by the fact that the definition of a "common law marriage" is itself vague. There are typically no symbols, such as a wedding licence, by which the existence of a common law marriage could be defined.

It is possible, however, to use census data to estimate the number of persons sharing their living quarters with an unrelated person of the opposite sex. This presumably includes many common law couples (but not those reporting themselves as "married") but might well include others as well. Nonetheless, it is the best indicator currently available. Three studies for the United States discuss such data: Carter and Glick (1976,

TABLE 8
Adult males (000s) by age and living arrangement, United States, 1975

	Under 35	*35–54*	*55 or older*
Married/spouse present	14,633	19,201	14,626
Living with unrelated person of opposite sex	531	183	171
Man never married	361	39	31
Man ever married	170	144	140

SOURCE: Glick and Spanier (1980, 23)

420-4), Glick and Norton (1977), and Glick and Spanier (1980). The data in Table 8 are taken from the last of these.

These data suggest that the extent of cohabitation was highest among persons under 35. There, including cohabitants as married would raise the number of "marrieds" by about 531/14,633, or under 4 per cent. In the older age groups, the cohabitants would add closer to 1 per cent to the number of marrieds. Such results are not large enough to account for the decline in first marriage rates. Remember that the FMR for 25-29-year-old males dropped from 0.20 in 1966 to 0.14 in 1981. Even after augmenting the 1981 FMR by a few percentage points to allow for common law marriage, the drop after 1966 was still substantial.

Marriage Swings and Fertility Swings

The postwar swings in fertility and first marriage are similar. To a large extent, the baby boom and marriage rush periods overlap as do the marriage bust and baby bust periods. Also, there are congruences in age patterns. In the baby boom, women undertook child bearing at younger ages compared to prewar years: they also married much younger. In the baby bust, women tended to have children at older ages: they also tended to marry later.

To an extent, this is not surprising. Existing cultural values emphasized marriage prior to child bearing. For many people, marriage was an important precondition to child birth. However, there need not be any especial connection between age at marriage and age at child birth. Throughout the postwar period, most women were married by their mid-twenties and, even given the increasing proclivity to complete child bearing by age 35, could plan births over the space of a decade or so. A shift in age at marriage of one or two years could, for most couples, be accommodated without necessarily altering decisions about the age sequencing of child birth.

Unfortunately, we know little about empirical relationships between marriage and child bearing for Canadian women. Until the early 1980s, this was also true of the United States. The work by Tsui (1982) on transition rates between birth parities is important in this respect. Tsui examined the fertility of once-married white U.S. women, still living with their spouses, in 1975.[4] She found that, at the peak of the baby boom, women marrying between 1955 and 1959 had a 26 per cent chance of first giving birth within two years of marriage and a 48 per cent chance of first giving birth within six years of marriage. Similar data for other marriage cohorts are given in Table 9. These data indicate that the cohorts married during the baby boom had a higher probability of early child birth (i.e. within two years of marriage) compared to cohorts married during either the prewar or baby bust period. In other words, the baby boom was characterized by a shortening of the interval until first child birth. Between 1945 and 1960 women were not only marrying earlier, they also had children sooner after marriage. Because of this, the typical age at first child birth for married women fell even faster than did age at first marriage.

From Tsui's data, it is possible also to calculate the probability of first child birth within six years of marriage. These probabilities, also shown in Table 9, indicate another way that the baby bust period differs from its predecessors. In the marriage cohorts up to 1940–5, the probability of child birth within six years hovered around 37 to 39 per cent. During the baby boom years, it rose to 43 to 48 per cent, then it fell back to prewar levels during the baby bust. In other words, women who married between 1965 and 1969 were still as likely as those who married during the baby boom to have had their first child within six years of marriage; they were just less likely to have had that child within two years. They were, however, more likely to have a child within the first six years than were women married in the prewar period. This indicates that women married during the baby bust were increasingly focusing their first births within an interval from two to six years after marriage.

Tsui also calculates transition rates between first and second (or higher-order) birth parities. These are also shown in Table 9. As in the 0–1 birth parity transition, the cohorts of women first married during the baby boom had an accelerated pace of child bearing compared to prewar cohorts. Also, like the 0–1 transition, the probability of a 1–2 transition within two years declined sharply, to prewar levels, for women marrying in the baby bust period. However, unlike the 0–1 transition, the probability of a 1–2 transition within six years also declined sharply for the baby bust cohort. This further illustrates that women marrying in the baby bust period spread out their second births over a longer interval. Unfortunately, Tsui's data, which cover child births only up to 1975, are insufficient to measure fully the extent of delayed second births in the baby bust cohorts.

TABLE 9
Transition rates between marital parities for once-married U.S. white women, still living with spouse, in 1975

	Marriage to first childbirth (0 to 1 birth parity)		*First to second childbirth (1 to 2 birth parity)*	
Marriage cohort	*Within 2 yrs*	*Within 6 yrs*	*Within 2 yrs*	*Within 6 yrs*
1930-4	0.167	0.370	0.050	0.189
1935-9	0.167	0.387	0.051	0.196
1940-4	0.157	0.388	0.054	0.229
1945-9	0.204	0.431	0.070	0.269
1950-4	0.202	0.462	0.095	0.325
1955-9	0.260	0.484	0.112	0.337
1960-4	0.230	0.450	0.094	0.304
1965-9	0.178	0.459	0.055	0.184

SOURCE: Computed from Tsui (1982, 7)

There was a close empirical relationship between the swings in fertility and in first marriage. To an extent, this is not surprising, since marriage was widely viewed as a precondition to child bearing. However, the sequencing of child birth for married women did change. During the baby boom, the pace of having children quickened. During the subsequent baby bust, this pace slowed as a new sequencing pattern took hold. In these important senses, the swings in fertility were not simply reflections of the swings in first marriage, even though the two behaved similarly.

DIVORCE, SEPARATION, WIDOWHOOD, AND REMARRIAGE

To understand fully changes in household formation, it is necessary to consider how families dissolve as well as form. This section considers dissolution arising from widowhood, separation, and divorce. Also considered is the regrouping of individuals through remarriage. Over the postwar period, there were substantial changes in each.

Separation and Divorce

Divorce, as with conventional marriage, has a legal definition. As such, it is relatively easy to measure. One can count the number of divorces granted in a given year and obtain information about the persons being divorced: e.g. current age, age at marriage, and number of dependent children. The same cannot be said of separations. Some separations do involve legal agreements between the spouses (as regards, for example, child support and visiting privileges), but these are essentially private documents for which there is no systematic reporting. Further, there need not be any legal agreement or other artifact to identify a separation. Consequently, there is little information on the extent of separation.

Divorce in Canada was less common prior to 1968. See McKie et al (1983, 19-54) for a historical review of divorce procedures in Canada. Between 1930 and 1947, the number of divorces in Canada went from 1,000 to 8,000 annually. In spite of this growth, the number of divorces never exceeded 2 per cent of the number of marriages. From 1948 to 1962, the annual number of divorces remained stable, at about 6,000. After 1962, the numbers climbed rapidly, spurting to 26,000 in 1969 and continuing to grow quickly through the 1970s. By 1981, there were 67,000 divorces, or more than one divorce for each three marriages.

In 1968, Parliament passed the Divorce Act. Prior to that, there were only two methods of obtaining a divorce. One was a "judicial divorce," obtainable only in provinces that had their own divorce legislation and only for certain types of marriage breakdown. The other was a "parliamentary divorce," literally a private act of Parliament. The Divorce Act of 1968 superseded existing provincial legislation. It redefined the bases for divorce and eliminated "parliamentary divorce."[5] The act liberalized divorce proceedings, helping to account for the subsequent rapid rise in number of divorces. One liberal feature of the legislation allowed, as a ground for divorce, simply separation, after three years.

What has been the impact on household formation of increasing marital disruption? The rapid rise in divorce suggests substantial effects on household formation, as former husbands and wives seek out different living arrangements. However, some of the rise in divorce in the postwar period meant that divorces replaced separations. If such persons make similar living arrangements, changing from separated to divorced may itself have little impact on household formation.

Although we lack regular, consistent historical information, the number of separations was apparently substantial. In the 1941 Census, almost six times as many persons were enumerated as separated compared to divorced (in this Census, unlike later ones, "divorced" includes persons who were legally separated). It was not until the 1976 Census that "separated" was again treated as a distinct marital status. Remember that, since this Census was self-enumerated (unlike 1941), some separated persons may have listed themselves incorrectly as "married." Nonetheless, McVey and Robinson (1981) note that the number of separated persons in 1976 was slightly larger than the number divorced. Thus, even after the liberalization of divorce in 1968, there were substantial numbers of separated persons. However, the ratio of separated to divorced persons fell sharply between 1941 and 1976, suggesting that part of the postwar rise in divorce was simply a substitution of divorce for separation.

TABLE 10
Age-specific divorce rates (%) for Canadian men and women, 1971-81

	Women			*Men*		
Age group	*1971*	*1976*	*1981*	*1971*	*1976*	*1981*
15-19	0.23	0.40	0.36	0.06	0.14	0.17
20-24	0.77	1.22	1.37	0.58	0.92	0.99
25-29	0.99	1.67	2.02	0.94	1.58	1.90
30-34	0.87	1.51	1.72	0.94	1.62	1.95
35-39	0.74	1.24	1.43	0.80	1.37	1.64
40-44	0.66	1.01	1.11	0.71	1.15	1.31
45-49	0.52	0.80	0.86	0.60	0.93	1.02
50 and older	0.24	0.35	0.32	0.30	0.43	0.42

SOURCE: Divorce data taken from *Vital Statistics, Vol 2: Marriages and Divorces*, various years. Married population data taken from census for respective year

One measure of the incidence of divorce is a divorce rate. This is, for a given cohort, the ratio of divorces in a given year to the number of married persons at mid-year.[6] Detailed data on divorces are unavailable prior to 1968. In Table 10 are shown age-specific divorce rates for Canadian men and women between 1971 and 1981. Corresponding data for the United States are reported in Carlson (1979).

These divorce rates help us better to understand why the number of divorces rose so quickly in the 1970s. In part, it is because the divorce rates themselves increased, particularly among 20-to-49-year-olds. In part, it is also because the baby boom cohort began to reach the prime ages for divorce. Among both men and women, divorce is most common between 25 and 34. By 1981, the baby boom generation had swollen the size of this age cohort.

What caused the rapid rise in divorce in the 1960s and 1970s? Several explanations have been offered, but few have been empirically tested. Preston and McDonald (1979) have estimated the probability of divorce for U.S. marriage cohorts since the Civil War. They find that this probability increased exponentially over time. Only about 6 per cent of U.S. marriages in 1870 ended in divorce. For 1920 marriages, this rose to about 18 per cent, and for 1960 marriages it is projected to be about 35 per cent. Given this trend, the postwar rise in divorce was not extraordinary.

Some argue that divorce is connected to fertility, insofar as the presence of children inhibits divorce. Such an argument is consistent with postwar evidence. As noted above, the proportion of a marriage cohort divorcing was below the long-term trend for cohorts marrying from 1945 to 1956, coinciding with the baby boom. Why does the presence of children inhibit divorce? One view is that children create additional bonds between the parents, which would have to be broken in the event of divorce. Cherlin (1977) looks at this argument using a large, longitudinal sample of U.S. women who in 1967 were aged 30 to 44, married, and living with a spouse. Cherlin followed this sample over the ensuing four years, identifiying those women who subsequently became separated or divorced (4.4 per cent of the sample). He found that the presence of young children (under 6 years of age) did inhibit the probability of divorce or separation during this period, but that the presence of older children did not. Cherlin (1977, 271-2) draws the following conclusion: "Children were a deterrent to divorce and separation only when they were in the preschool ages, when the time and effort required for child care are at their peak. These findings suggest that children prevent marital dissolution not because they build new bonds between parents but rather because early child care may be too expensive and time-consuming for one spouse to manage alone."

The reverse argument has also been made, that impending divorce or separation serves to reduce fertility. Typically, the incidence of child birth falls before and during a period of separation or divorce. Thornton (1978) presents evidence of lower fertility during the two years before a separation or divorce. This Thornton attributes to the marital discord that typically precedes dissolution. However, he also finds evidence of "make up" fertility among women who eventually remarry; they tended to have roughly the same completed level of child bearing as other married women. In other words, lower fertility during separation and divorce was made up during the early years of remarriage.

Finally, others argue that it is unwise to look for connections between divorce and fertility at all. Preston and McDonald (1979), for example, deliberately ignore the impact of fertility on divorce in their empirical work. They comment (15): "The congruence between marital cohort divorce and marital cohort fertility is striking. Marriages contracted just before and during the depression were characterized by low fertility and those contracted after the war and through the mid-fifties were relatively fertile . . . Interpretation of this relation is difficult since marital dissatisfaction can manifest itself both in low fertility and in high instability. It seems worthwhile to attempt to locate factors that are causally prior to both, recognizing that fertility and stability are in a sense joint outcomes."

Widowhood

For men, the general postwar increase in longevity was manifested largely in declining mortality under age 60. For women, there were substantial reductions in mortality rates both above and below age 60. Census data attest to the declining incidence of widowhood among persons aged under 60. For example, 17 per cent of Canadian women aged 55 to 59 were identified as widowed in 1951, compared to 16 per cent in 1961, 14 per cent in 1971, and 13 per cent in 1981. However, censuses recorded only the current marital status; widowed persons who remarry are identified as "married." Because the incidence of remarriage varied widely over the postwar period, census counts of the currently widowed do not correctly indicate the extent or incidence of widowhood.

Also, the above-noted mortality trends do not distinguish among marital statuses. The only source of marital-specific mortality rates for Canada during the postwar period is the marital status life table in Adams and Nagnur (1981). Their table, based on data for 1975-7, shows that mortality rates for men, for example, did vary considerably by marital state. Men marrying exactly once had the greatest life expectancy. Such marital status differences make it difficult to generalize about the entire postwar period in the absence of any earlier marital status life tables. To what extent, for example, was the postwar increase in longevity of men attributable to a higher propensity to marry during the "marriage rush"? Did mortality rates among men conversely rise as a result of the post-1960 "marriage bust"?

However, even in the absence of adequate data on the incidence of widowhood, it is apparent that the number of elderly widows increased rapidly. As an example, there were just 48,000 widows aged 70 to 74 in the 1941 Census. This increased to 155,000 by 1981. The swelling number of elderly widows was an important component of total household formation in Canada during the postwar period.

Remarriage

When a marriage is dissolved through death or divorce, surviving partners may choose to remarry. Remarriage is very common. In fact, for as far back as such vital statistics are published (roughly about 1940 in Canada), divorced or widowed persons have been generally more likely to marry than have been single persons of the same age. Adams and Nagnur (1981, 67-70) point out that in 1975-7 a single Canadian male at age 30 had an 11 per cent chance of marrying within the following year. For a widowed man of the same age, the probability of marriage was 19 per cent, and for a divorced man, almost 44 per cent.

One measure of the extent of remarriage is the widowhood remarriage rate (WRR) or the divorced remarriage rate (DRR), formed by dividing the number of remarriages of widowed (or divorced) persons during a year by the Census count of widowed (or divorced) persons at mid-year. These are similar, in construction, to first marriage rates.[7] Basavarajappa (1978, 32) presents WRR and DRR for Canadian men and women between

TABLE 11
Widowhood and divorced remarriage rates for selected cohorts, Canada, 1941-81

Marital status	*Sex*	*Age*	*1941*	*1951*	*1956*	*1961*	*1966*	*1971*	*1976*	*1981*
Widowed	Men	30-34	0.147	0.232	0.203	0.162	0.204	0.103	0.176	0.159
	Women	25-29	0.085	0.171	0.165	0.130	0.128	0.096	0.109	0.109
Divorced	Men	30-34	0.417	0.716	0.762	0.697	0.656	0.353	0.363	0.296
	Women	25-29	0.348	0.479	0.564	0.456	0.402	0.286	0.262	0.246

SOURCE: Computed from vital statistics and census data, various years

1941 and 1971. After controlling for age and sex, the marriage rates for divorced persons are generally higher than for widowed persons, and both are higher than the corresponding FMR. These rates are also much higher for men than women: widowed or divorced women were much less likely to remarry than men at the same age. Finally, as with first marriage rates, the DRR and WRR drop quickly with age; older men and women are less likely to remarry than their younger counterparts.

Earlier in this chapter, "marriage boom" and "marriage bust" periods were identified on the basis of changes in FMR. Do DRR and WRR display the same temporal pattern? Consider the data presented in Table 11 for two representative cohorts: men aged 30-34 and women aged 25-29. They are quite variable. There is evidence of an abrupt drop after 1971, corresponding to the "marriage bust." However, among widowed men, remarriage rates rose again in 1976 and 1981. As well, remarriage rates did not remain uniformly high during the "marriage rush" period. Among both men and women, WRR subsided after 1951; DRR peaked closer to 1956. What generated these changes in remarriage is not clear. What is clear, however, is that they behaved differently from FMR.

Glick and Norton (1977) note a similar decline in remarriage propensities in the United States. They note also that the typical duration of a first marriage shortened, as did the time between divorce and remarriage. This implies a compression of marital events as individuals moved more quickly from marriage to divorce to remarriage. This compression reduced the effect of marital instability on net household formation.

FAMILY FORMATION

In chapter two, a nuclear family is defined to be a husband-wife couple coresiding with or without any never-married children or a lone parent coresiding with one or more never-married children. The formation of families is of especial interest to us because the large majority of households in postwar Canada consisted of such families living alone.

Part of the story of postwar family formation is evident from the above review of marriage and fertility trends. For the most part, families are created and change through marriage, child birth, divorce, and widowhood. The rise in typical family size during the baby boom period, its decline during the baby bust, and the rise of the lone-parent family during the marriage bust period are explicable in these terms to varying degrees.

However, some aspects of family formation and change are not evident from fertility and marriage data alone. First there is the incidence of common law marriage. Census counts of families largely do not distinguish between conventional and common law marriages. However, the vital statistics data, presented in the previous two sections, are for conventional marriages only. In this sense, vital statistics data on marriages are of limited assistance in the analysis of Census data on husband-wife family formation.

Second, the fertility data presented to this point do not distinguish between child birth inside and outside marriage. It is not easy, therefore, to determine the extent of lone-parent family formation arising from child birth outside marriage.

Third, there is the problem of assessing family change arising because of changes in the number of children present in a family. At best, fertility data give us an idea of "completed" family size: i.e. the total number of children borne by a mother. Some of these children may die before reaching adulthood. Others marry and, by definition, no longer form part of their parents' nuclear family, regardless of whether they coreside with their parents. Still others leave a parental home to live on their own and thereby no longer form part of the parents' nuclear family. The formation of a family has to be viewed in its dynamic setting, tracing the arrival of new children and their eventual marriage, home leaving, or death. This temporal sequencing differentiates actual family size, at any age, from "completed" family sizes evidenced in fertility data.

These problems make it useful to analyse family formation and change separately, as well as from the perspectives of fertility and marriage. In the two succeeding subsections, evidence from various censuses of nuclear family formation and change is considered.

Family Formation by Type

During the marriage rush, marriage was both earlier and more widespread. In its early phase, the marriage rush was accompanied also by a slowdown in the growth rate of divorce. Thus there was a rapid growth of husband-wife families through the 1950s, even though the number of young adults grew only modestly. However, during the subsequent marriage bust, the frequency of first marriage declined and the incidence of

TABLE 12
Nuclear families (000s) by type, Canada, 1941–81

Type of family	*1941*	*1951*	*1956*	*1961*	*1966*	*1971*	*1976*	*1981*
Husband-wife	2,203	2,962	3,393	3,800	4,154	4,605	5,175	5,611
Lone parent								
Male parent	81	75	75	75	72	100	95	124
Female parent	226	251	243	272	300	371	464	590
Total	307	326	318	347	372	470	559	714
All families	2,510	3,287	3,712	4,147	4,526	5,076	5,735	6,325

SOURCE: Taken from the census of Canada, various years

divorce rose. These dampening effects were offset, though, by the swelling number of young adults (i.e. the aging baby boomers). Consequently, the number of husband-wife families continued to grow briskly throughout the 1960s and 1970s.

What happened during these periods to the formation of lone-parent families? Here, the answer is not as evident from the preceding discussion. Unless divorce was simply replacing separation, the rising incidence of divorce should have led to more lone-parent families. However, divorce rose as fertility fell (and the incidence of childlessness rose); the divorce of a childless couple does not result in a lone-parent family. Also, the continued decline in mortality among under-60-year-olds during the postwar period reduced the number of lone-parent families formed by widowhood. The net effect of these various processes on the formation of lone-parent families is unclear, a priori.

In Table 12 are presented some counts of families by type. Census data prior to 1941 are not comparable because they include in the family other relatives, such as grandparents and uncles. Remember also that the definition of a family was slightly different in the 1941 Census, because it assumed coresidency within a housekeeping unit, which could be smaller than the household as defined in this book.

These data show the relatively quick growth of husband-wife families, up to about 1961. A slowdown during the early 1960s was followed by more rapid growth in the 1970s. Overall, the number of husband-wife families grew at 2.2 per cent annually between 1951 and 1981 (compared to 1.9 per cent for population). The pattern among lone-parent families was different. The number of lone-parent families remained constant during the 1950s, then rose sharply. The growth rate is especially marked for female lone-parents after 1966, rising at an average 4.6 per cent annually, compared to 1.3 per cent for population over the same period. Similar findings for the United States are described in Glick and Norton (1977, 27).

TABLE 13
Ratio of husband-wife families to population of married males by age group of male/husband, Canada, 1951-81

Age group	*1951*	*1956*	*1961*	*1966*	*1971*	*1976*	*1981*
15-24			0.929	0.938	0.928	0.929	0.931
25-34	0.955*	0.953*	0.960	0.965	0.950	0.945	0.942
35-44	0.955	0.957	0.956	0.959	0.949	0.953	0.948
45-54	0.936	0.945	0.949	0.957	0.944	0.951	0.946
55-64	0.924	0.932	0.934	0.947	0.940	0.950	0.948
65 or older	0.918	0.922	0.905	0.922	0.917	0.926	0.924

SOURCE: Computed from census of Canada, various years
* In 1951 and 1956, separate data are not available for 15 to 24-year-olds and 25 to 34-year-olds. Data shown are actually for the 15-34 age group.

Overall, these patterns coincide with the marriage rush and bust periods. They also show the effects of the baby boom cohort as it began to reach the family formation ages in the late 1960s. The rise of lone parenthood in the 1960s and 1970s is not surprising, given high divorce rates. However, the impact was ameliorated by the rising incidence of childlessness.

To control for the baby boom and other age composition effects, one can disaggregate these data by age groups and then divide by some relevant control population. In Table 13, the number of husband-wife families in a particular age cohort of husband is divided by the number of married men of that age. For example, on average 931 of 1,000 married men aged 15-24 in 1981 were living in husband-wife families. The remaining 79 were living apart from their spouses. Some were separated, either legally or informally, while others were temporarily living apart (e.g. because of job commitments).

These ratios are all close to 1.0, indicating that nearly all married men lived in husband-wife families. There was a slight upward trend over the marriage rush period, indicating that a rising portion of marrieds lived with their spouses. Put another way, the changes between 1951 and 1966 were consistent with a slight decline in the incidence of separation. After 1966, these ratios slumped among the under-55 age groups in general, and among 25-34-year-olds in particular. This is consistent with the increasing incidence of separation that accompanied rising divorce during the 1970s.

These data shed light on the role of substitution between divorce and separation. That married persons may be living apart from their spouses for reasons other than marital incompatibility limits the extent to which such ratios can be used to analyse separation. Nonetheless, the ratios were relatively stable over the postwar period. If divorce had simply been a widespread substitute for separation after the mid-1960s, we should have seen corresponding increases in these ratios. However, if anything, the ratios slumped in the 1970s. This suggests that divorce was not a broad substitute for separation, that overall marital instability (separations plus divorces) increased as divorce rates climbed.

Earlier in this chapter, it is noted that after 1968 it became possible to obtain a divorce after three years of separation. Separation, in conjunction with irreconcilable differences, became a popular ground for divorce. This helps to explain why the incidence of separation did not decline as the incidence of divorce rose.

A similar analysis is possible with lone-parent families. Table 14 shows the ratio of lone-parent families to the adult population of that age and sex. Let us refer to such ratios as lone parent-family headship rates (LPFHRs). For example, on average 95 of every 1,000 women (regardless of marital status) aged 35-44 headed a lone-parent family in 1981, for an LPFHR of 0.095.

In examining this table, note that our definition of a family says nothing about the ages of family members. A lone-parent family could, for example, consist of a 70-year-old woman and her 40-year-old daughter. To be a "child," one need only be never married and living with a parent. This helps to explain why there are so many lone-parent families headed by individuals aged 65 or older. A conventional image of a lone-parent is a young to middle-aged adult with dependent children. The definition of a family employed here includes these families as well as those in which the "child" is an adult, possibly even supporting the parent.

In table 14 opposite, the LPFHRs for persons aged 55 or older are substantially different from the younger age groups. They declined steadily from 1951 to 1981. Why? Most children leave a parental home as young adults. When all the children in a lone-parent family have left home, the family (as defined here) no longer exists; the lone-parent becomes a non-family person. The age at which that happens varies considerably; it depends both on the parent's age when the children are born and on the ages at which these children leave home. In general, the earlier one has children, the fewer children one has, and the younger they leave home, the sooner a lone-parent family will cease to exist. This introduces two possible reasons for the decline in lone parenthood among those over 55.

One explanation is based on changing fertility. The fewer children a parent has during his or her lifetime, the less likely they are still to have a child at home after age 55. Remember that the cohorts that produced declining fertility in the 1920s and 1930s reached age 55 in the 1950s and 1960s. During the 1950s and 1960s, the LPFHR fell correspondingly. However, the cohorts that produced the postwar baby boom reached age 55 starting only in the late 1970s. Note the corresponding, small upturn in the LPFHR in 1981.

TABLE 14
Ratio of lone-parent familites to adult population by age group, Canada, 1951-81

Age group	*1951*	*1956*	*1961*	*1966*	*1971*	*1976*	*1981*
(a) Men							
15-24			0.001	0.001	0.002	0.001	0.001
25-34	0.002*	0.003*	0.005	0.004	0.012	0.007	0.007
35-44	0.011	0.010	0.010	0.010	0.017	0.016	0.021
45-54	0.021	0.018	0.016	0.015	0.020	0.020	0.028
55-64	0.030	0.025	0.021	0.018	0.018	0.017	0.022
65 or older	0.050	0.044	0.039	0.031	0.023	0.020	0.018
(b) Women							
15-24			0.008	0.008	0.012	0.015	0.020
25-34	0.019*	0.014*	0.025	0.029	0.046	0.055	0.066
35-44	0.049	0.041	0.044	0.047	0.062	0.078	0.095
45-54	0.074	0.063	0.065	0.066	0.072	0.080	0.092
55-64	0.095	0.083	0.073	0.069	0.066	0.066	0.070
65 or older	0.124	0.112	0.100	0.089	0.065	0.060	0.053

SOURCE: Computed from census of Canada, various years
* See note to Table 13.

Another explanation focuses on changes in home leaving. There has always been much variation in age at home leaving. For a variety of individual, social, psychological, and economic reasons, some young adults leave home much earlier than others. However, the postwar period was marked by a more widespread home leaving at the younger adult ages. That earlier home leaving reduced the likelihood that someone would still be a lone parent at or after age 55.

Let us turn now to the under-55 age groups. There were relatively few male lone parents; note the low ratios in panel (a). Most lone parents were women, and, as seen in panel (b), their LPFHR increased rapidly during the postwar period. What caused this rapid rise in female lone parenthood? In large part, it reflects the impact of rising divorce in the 1960s and 1970s. Although there was an upturn in LPFHR among men in the late 1970s, as late as 1981 women were still about four times as likely as men to be lone parents.

The emergence of the lone-parent family was not unique to Canada. Glick (1976, 321-3) notes that, in the United States, the proportion of children under 18 who were living with just one parent almost doubled between 1960 and 1974. In 1974, over 90 per cent of such children were living with the mother. Glick notes that only two children of three on average were living with both parents in an intact first marriage.

However, it is easy to overemphasize the importance of lone-parent family formation. The decline in childlessness and falling mortality rates meant that, in 1980, more women were surviving to the end of their child bearing years, having borne at least one child and being in an intact first

TABLE 15
Percentage distribution by family status at age 50 of U.S. white women reaching age 15, by birth cohort

	Women by year of birth			
	1900–4	*1910–14*	*1920–4*	*1930–4*
Died before age 50	12.5	9.0	6.0	5.0
Never married at age 50	7.5	6.0	5.0	4.5
Ever married at age 50				
Childless	15.5	14.5	8.5	5.5
Ever had children				
Still in first marriage	47.5	52.0	60.5	64.5
Not in first marriage	17.0	18.5	20.0	20.5
All women at age 15	100.0	100.0	100.0	100.0

SOURCE: Uhlenberg (1974, 286)

marriage. Uhlenberg (1974) illustrates this for U.S. white women born up to 1934 who survived to at least age 15. See Table 15. Uhlenberg's data show the rising incidence of marital dissolution; the proportion of women whoe survive to age 50, have children, but are not in an intact first marriage, rose from 17 per cent for the 1900–4 birth cohort to 20 per cent for the 1930–4 cohort. In part, this illustrates the trend to lone parenthood. However, during the same time, the proportion having children and still in an intact first marriage at age 50 rose from 48 to 64 per cent. Why? Declining mortality under age fifty (especially for men), and the increased propensity to marry and to have children observed during the first half of this century, overwhelmed the trend to divorce and separation.

Family Formation by Size

A dominant characteristic of postwar Canadian household formation has been the shrinking average size of household. After about 1966, this smaller number of persons per household was in part attributable to the shrinking size of an average family. Average family size increased slightly, from 3.7 persons in 1951 to 3.9 in 1966, then fell rapidly, down to 3.3 persons by 1981.

In a sense, this is not surprising. We have already noted the extensive changes in postwar fertility, first in the "baby boom" and then in the "baby bust" period. These changes undoubtedly affected family size. Other things being equal, typical family size should have increased during the baby boom and fallen during the baby bust. A completed average family size, as defined above, is simply the number of live births typically

TABLE 16
Completed family size by birth cohort of prospective mother, U.S. white women born between 1880 and 1929

Birth cohort of mother	Completed family size 0	1	2	3	4	5	6	7+
1880-4	0.218	0.130	0.145	0.121	0.096	0.074	0.059	0.156
1885-9	0.212	0.142	0.157	0.130	0.099	0.071	0.056	0.133
1890-4	0.195	0.161	0.180	0.140	0.100	0.066	0.049	0.109
1895-9	0.190	0.188	0.203	0.142	0.091	0.057	0.042	0.086
1900-4	0.198	0.214	0.215	0.137	0.083	0.050	0.035	0.067
1905-9	0.207	0.220	0.230	0.137	0.080	0.045	0.029	0.052
1910-4	0.188	0.208	0.249	0.154	0.085	0.045	0.027	0.044
1915-9	0.144	0.181	0.267	0.180	0.102	0.053	0.029	0.043
1920-4	0.099	0.157	0.268	0.205	0.124	0.065	0.035	0.048
1925-9	0.093	0.121	0.245	0.220	0.144	0.079	0.043	0.054

SOURCE: Taken from Heuser (1976, 237)

experienced by women in a cohort. Because of death, home leaving, or marriage, some of these children may not be enumerated as part of their parents' family on a given date. Nonetheless, changes in completed average family size are a useful indicator of the role of fertility in shaping observed family sizes.

In the absence of Canadian data, let us consider the estimates of Heuser (1976) for U.S. white women. In Table 16 are presented his distributions for women aged 45-49, virtually at the end of their child bearing years. Constrained by the fact that one must wait until a cohort has completed its child bearing before one measures completed family sizes, Heuser was limited to cohorts born before 1930. Shown in this table are the cohorts that produced the declining fertility of the first part of this century and the subsequent baby boom. The cohorts that produced the post-1960 baby bust were too young to have fully completed child bearing by the time of Heuser's study. Nonetheless, these data provide valuable insights. In general, note the rising popularity of the two- and three-child family. In the 1880-4 birth cohort, just 27 per cent of women had two or three children. However, for women born between 1925 and 1929, this rose to 46 per cent. Note also the increased frequency of lone-child families among women born in the first decade of this century (many of whom had their child in the Depression. Note as well the heightened frequency of 3+ children families among women born in the 1920s (the mothers of the early baby boom years).

In contrasting these completed family sizes with census data on family size, remember that fertility decisions, for older families, were typically made some time before. For example, a 45-year-old woman in 1971 typically bore her children in the 1950s; her current family size typically reflects earlier fertility decisions. In the same way, as the mothers of the baby bust reach age 45, their typical family size will reflect past fertility.

In addition, average family size can change even in the absence of fertility changes. This is because an average family size is derived from a "snapshot," such as a Census, of a set of families at one point in time. As such it includes some young families, whose child bearing may not even have begun, some families who have completed child bearing but not yet experienced any home leaving, and some older "empty nester" families with no children remaining in the parental home. The average number of persons in a family will vary with the mix of these types of families. In other words, average family size is sensitive to the age distribution of the adult population.

Average family size changed for other reasons as well. One of these was the variation in spacing of children. We have already noted that the baby boom period was marked by a reduction in the average age of mothers at both first and last child births, a reduction in the incidence of childlessness, and an increase in the incidence of higher-order births (notably third births). However this implies that young couples were more likely to have at least one child, more children on average, and completed child bearing sooner. As these same couples age, their children reach home leaving age correspondingly sooner in terms of the parents' ages. In the baby bust period, child bearing was initiated later, fewer children were borne, and the last child was borne at a typically younger age. The baby bust can thus be characterized as a "bunching up" of a smaller number of child births within a narrow range of ages of mother. This reduced the range of ages at which the mother has all her children present.

Figure 5 is illustrative. Panel (a) shows a life cycle for a woman who might be a typical mother of the baby boom. She had three children: labelled A, B, and C. A was born when she was 23, B at age 27, and C at 35. I assume that each leaves home at age 21. The number of children present each year is indicated under the diagram. Note that all three children are present in the family for only 9 years (when the mother is 35 through 44), but the last child does not leave until the mother is 56. Panel (b) shows a life cycle diagram for a mother typical of the baby bust. She has just two children (A and B). The first is born when she is 26 (3 years later than the baby boom mother) and the second when she is 29. All her children are present in the family for 18 years, much longer than for the baby boom mother. However, her last child leaves when she is just 50.

Let us now examine changes in average family sizes as reported in various censuses in the postwar period and summarized in Table 17. Consider first the husband-wife family data in panel (a). The postwar baby boom shows up most clearly in the younger age groups. For example, among families wherein the husband was 35–44, the average family size

FIGURE 5

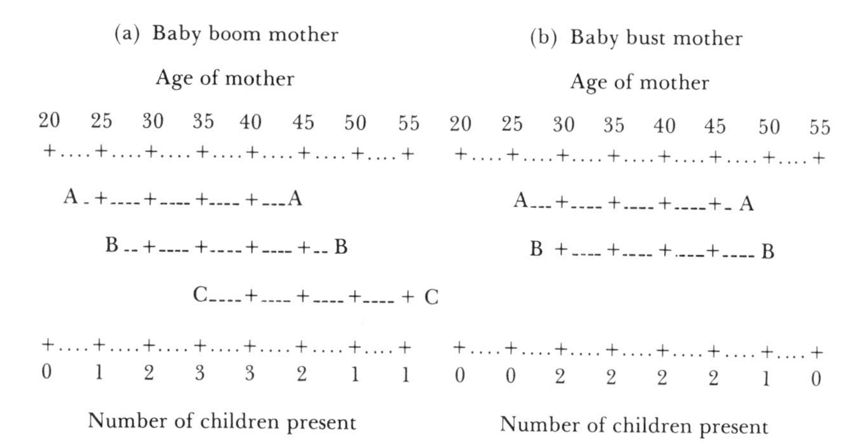

(including the two spouses) increased from 4.5 persons in 1951 to 5.0 persons in 1966. Above age 44, the postwar baby boom had little noticeable effect, since earlier and more prevalent home leaving offset higher completed family sizes. After 1961, the effect of the baby bust became apparent, first in the under-25 group and later at older ages. Remember, though, that as late as 1976, the 35-44 age group still included many wives who initially bore children during the late baby boom period. The 1981 Census data show a continued sharp downturn in average family size in this age group. At the same time, the decline in family sizes among the over-44 age group reflects the continuing trend to youthful home leaving.

Panel (b) gives average sizes for female lone-parent families. These data suggest that the lone-parent family grew during the baby boom and shrank during the baby bust, much as did the husband-wife family. The average size of a female lone-parent family was, if anything, slightly less sensitive to variations in fertility rates.

The Canadian and American experiences are broadly similar in this respect. In a comparative study, Rodgers and Witney (1981) indicate the similarities but also point out a few important differences. Canadians tended to first marry a little later than Americans (the gap was much larger at the turn of the century but diminished during the postwar period). Canadian women also tended to have their first child later but to space their child births closer together, with the result that they were typically younger at completion of child bearing. When these factors are combined with higher life expectancy, Canadian families typically experience a longer empty nester period. Thus, typical family sizes in Canada, when averaged over all age groups, should be smaller than in the United States, in the absence of other differentials in completed family size.

TABLE 17
Average size of husband-wife family and female lone-parent family by age of husband or mother, Canada, 1951-81

	1951	*1956*	*1961*	*1966*	*1971*	*1976*	*1981*
*(a) Husband-wife families: age of husband**							
Under 25			2.9	2.8	2.6	2.5	
25-34†	3.5	3.7	4.0	4.0	3.7	3.4	3.3
35-44	4.5	4.7	4.8	5.0	4.9	4.5	4.2
45-54	4.3	4.3	4.3	4.4	4.3	4.1	3.5
55-64‡	3.3	3.3	3.2	3.2	3.1	2.9	2.4
65 or older	2.6	2.5	2.4	2.4	2.3	2.3	
(b) Female lone parents: age of mother							
Under 25			2.6	2.6	2.5	2.4	
25-34†	2.9	3.0	3.3	3.5	3.3	2.9	2.6
35-44	3.4	3.4	3.5	3.7	3.8	3.5	3.1
45-54	3.3	3.2	3.1	3.2	3.2	3.1	2.9
55-64‡	2.8	2.8	2.7	2.6	2.6	2.5	2.3
65 or older	2.4	2.4	2.4	2.3	2.4	2.2	

SOURCE: Data taken from the census of Canada, various years
* In the 1981 Census, average sizes of husband-wife families shown by age of wife
† In the 1951, 1956, and 1981 censuses, the youngest age group published is "Under 35." These data are shown in the rows "25-34." For comparison, the 1961 mean husband-wife family size among under-35s was 3.8; for female lone parents, it was 3.1.
‡ In the 1981 Census, oldest age category given is "55 or Older."

CONCLUSIONS

This chapter examines postwar trends in fertility, mortality, migration, marriage, and marital dissolution. All of these trends had implications for the number of families and typical family sizes. In addition, a distinction is drawn between average family size (the number of family members present in a household at any one time) and completed family size (the number of children ever born). Average family size reflects the age distribution of the population, the spacing and number of child births, and the typical age at home leaving. Postwar changes in all of these help us to understand how household formation was transformed.

The chapter begins with a review of postwar population growth. Changes in the timing and number of births are noted. In the baby boom, the trend was to earlier, and more widespread, child births. Especially noteworthy was the rising incidence of women experiencing three births. In the baby bust (the 1960s and 1970s), the trend was toward fewer third births, with first births occurring later than before and the last before age 35. The incidence of childlessness, which dipped below 10 per cent in the baby boom, rose again in the baby bust period. Changes in mortality were also important. With increasing longevity, Canada's population grew. At the same time, this meant relatively more older families (especially empty nester couples), hence a smaller average family size. Also, the growing differential between women and men meant a rapidly increasing number of elderly widows. Another important determinant of postwar population growth was net immigration. Although the role of immigration varied considerably over time, it became an increasingly important source of growth with the post-1960 decline in fertility.

The chapter also considers changing nuptiality. It notes the tendency toward earlier and more widespread marriage during the marriage rush. After about 1970, however, these patterns began to revert to those of the prewar period, with a growing incidence of delayed first marriage and a declining incidence of first marriage overall. The rising incidence of divorce is also noted.

The effects of these different sources of population growth and marital status change on family formation are subsequently discussed. The rapid formation of female lone-parent families and the changing average sizes of husband-wife families in the postwar period can be explained partly in these terms. At the same time, family formation and size reflect also changes in home leaving.

This chapter illustrates the many different trends that helped shape the number, size, and composition of postwar families. It also discusses some of the principal explanations of these trends. From this discussion, it is evident that a variety of economic and social forces underlie the observed postwar demographic shifts, and the changes in family formation that ensued. Already, we can begin to see the futility of attempting to rely on a single, simple explanation of changing living arrangements. To argue, for example, that postwar family formation can be analysed simply in terms of demographic variables is to risk ignoring the important roles that economic forces played in shaping these.

Further, this chapter emphasizes how family formation is itself in part a reflection of living arrangement choices. The home leaving of young adults affects family size and can even lead directly to family dissolution. Also, separation and divorce inherently involve a living arrangement choice: to live apart from one's former spouse. In the next chapter, we consider how changes in family formation reshaped living arrangements. Although this approach is insightful, we need to be mindful of the fact that changing living arrangements in part helped reshape family formation.

CHAPTER FOUR

Postwar Household Formation

Having reviewed post-Second World War patterns of demographic change and their implications for family formation, we are now ready to look at living arrangements. This chapter begins by comparing the growth of population, families, and households. In Canada, as in other countries, the growth of households was much more rapid than either population or families. The number of households increased faster than population, and average household size declined. The remainder of this chapter considers possible explanations.

In a second part of the chapter, postwar change in the size composition of households is examined. The rapid growth in persons living alone (i.e. one-person households) and the decline of large households are highlighted. Looking at changes in size distribution helps in identifying important contributing factors.

In part, changing household size reflected fertility and family size patterns. Certainly, the post-1960 decline in household size occurred concurrently with the baby bust. However, average household size declined even during the baby boom, when families became larger. The contradiction between larger families and smaller households during the baby boom is assessed in the third section of this chapter.

A fourth section concerns the effects of age composition on household size. Average household size may decline simply because there are more young or old adults (who typically live in smaller households) and fewer middle aged (who more often head larger households). This section examines the extent to which postwar changes in average household size were due to such age composition changes. The period after 1960 is described in chapter three as the "marriage bust," because of declining nuptiality. Given that unmarried adults typically live alone or in other small nonfamily households, the marriage boom and bust affected household formation and average size. Their effects are also assessed in the fourth section of this chapter.

The prevailing method of analysing household formation employs statistics called "total headship rates." In a fifth section of this chapter, these statistics are defined and postwar patterns are described. After arguing that total headship rates are not a satisfactory tool, I show that the propensities of family units and nonfamily individuals to live alone provide more insight into household formation. This idea is pursued further in the two subsequent sections of the chapter, leading up to the components of change analysis in chapter five.

The final part of the chapter describes (longer-term) historical trends in average household size. Historical analyses are hampered by a lack of comparability among different definitions of households and dwellings. However, roughly comparable data from different time periods indicate a decline in average household size, possibly back into the nineteenth century. Some estimates of average household size in Europe, from as early as the seventeenth century, suggest that households typically never were very large and that there is reason to believe that the decline in average household size in Canada began early in the twentieth century.

AVERAGE HOUSEHOLD SIZE IN THE POSTWAR PERIOD

Between 1951 and 1981, Canada's population grew briskly. The average annual growth rate was 1.9 per cent. At that rate, population would have doubled in just 38 years.[1] The right hand side of Table 18 shows numbers of private households. This excludes persons living in collective dwellings such as residences, hotels, hospitals, and penal institutions. The number of private households increased by an average of 3.0 per cent annually between 1951 and 1981, for a doubling time of 23 years. In the time that the population of Canada takes to double at the 1951–81 average, the number of households would treble.

Between 1951 and 1981, the number of families grew at 2.2 per cent annually. That this was more rapid than the growth in overall population helps to explain why household formation proceeded so quickly. However, it is not the entire story. Households formed even more rapidly than did families.

Finally, Table 18 shows also the average number of persons per private dwelling. Over the postwar period, the average size of household declined steadily. By 1981, 100 private, occupied dwellings contained typically just 287 persons. This is the lowest average size ever recorded in Canada. Note that average household size fell even between 1951 and 1966, years during which completed family sizes were increasing. A similar pattern occurred in the United States. Average household size there fell from 3.5 persons in 1950 to 2.7 persons in 1981.[2] Estimates for dates in between

TABLE 18
Population, families, and households in Canada, 1951-81

	Population				*Families**		*Private households*		
			Private						
Year	*Total (000s)*	*Collective*† *(000s)*	*Total (000s)*	*Growth rate*‡ *(%)*	*Total (000s)*	*Growth rate (%)*	*Total (000s)*	*Growth rate (%)*	*Avge size*
1951	14,009	437	13,572		3,287		3,409		3.98
1956	16,081	633	15,448	2.6	3,712	2.5	3,924	2.9	3.94
1961	18,238	626	17,612	2.7	4,147	2.2	4,555	3.0	3.87
1966	20,015	609	19,406	2.0	4,526	1.8	5,180	2.6	3.75
1971	21,568	535	21,034	1.6	5,071	2.3	6,041	3.1	3.48
1976	22,993	581	22,412	1.3	5,728		7,166	3.5	3.13
1981	24,343	546	23,797	1.2	6,324	2.0	8,282	2.9	2.87

SOURCE: Computed from published census reports, various years

* Data are census counts of families. These include only husband-wife and lone-parent families. See chapter two for more explanation. The 1976 and 1981 counts for families include only families living in private dwellings; the small numbers of families living in collective dwellings are excluded. Prior to 1976, families living in collective dwellings are included. For this reason, the annual growth rate between 1971 and 1976 is not shown.

† In Canadian censuses, dwellings of a commercial, institutional, or communal nature are termed collectives. Persons residing in such dwellings form the collective population. All other persons reside in private dwellings and are termed private individuals. See chapter two for more detail.

‡ Growth rate shown is the annual compounded rate of increase over the preceding five-year period.

show a steady decline, similar to that for Canada. Kobrin (1976a) generated historical data on average household size for a consistent definition of household. Changes in household definition limited Kobrin to comparisons of the 1790, 1900, 1950, and more recent censuses; nonetheless, she found evidence of a systematic decline in mean household size since the earliest census.

THE SIZE COMPOSITION OF HOUSEHOLDS

Accompanying the decline in average size were important changes in the size composition of households. These are summarized in Table 19. Family sizes increased through the early 1960s as a consequence of the 1950s baby boom and thereafter declined. The fertility decline is reflected in the household size distributions after 1961. Note the declining proportion of households containing more than four persons - slowly between 1961 and 1966 and more rapidly thereafter. The pre-1961 data do not indicate

TABLE 19
Percentage distribution of households by size, Canada, 1941–81

Persons in household	*1941*	*1951*	*1956*	*1961*	*1966*	*1971*	*1976*•	*1981*
1	6	7	8	9	11	13	17	20
2	18	21	22	22	23	25	28	29
3	20	20	19	18	17	17	18	18
4	18	19	19	18	18	18	18	19
5	13	13	13	13	13	12	10	9
6	9	8	8	8	8	7	5	4
7	6	5	5	5	4	4	2	1
8 or more	10	7	7	6	6	4	2	1
Total	100*	100	100	100	100	100	100	100

SOURCE: Computed from Wargon (1979a, 43) and census data for various years
* Columns may not add due to rounding.

a buildup of large households, as might have been expected during the baby boom. Over the entire postwar period, the number of one-person households increased dramatically. To a lesser extent, so did the number of two-person units. In relative terms, the number of three-person units fell sharply between 1951 and 1961 and subsequently stabilized, while four-person households remained remarkably constant at about 18 per cent of all households.

These patterns require some explanation. Why, for example, in the "marriage rush" of the 1950s was there a rising incidence of persons living alone? Why during the baby boom did two-person units become more common? Why did the percentage of households that are large not increase during the baby boom?

THE FAMILY COMPOSITION OF HOUSEHOLDS

In looking at why average household size did not keep pace with growing families, it is helpful to disaggregate by family composition. Seven of ten postwar households were made up of families living alone; in chapter two, these are termed PNF households. Since average family size increased up until about 1966, presumably so too did the size of a typical family living alone. However, one must be careful here, since a family can live alone, maintain its own dwelling but have other persons also living with it, or lodge with another family or individual who is maintaining the dwelling. Of these, lodging families tended to be small (often just a husband-wife pair), very young or elderly, and residing with another set of relatives (often parents or children). Because lodging families were typi-

TABLE 20
Average size of family and PNF household, Canada, 1951–81

	1951	*1956*	*1961*	*1966*	*1971*	*1976*	*1981*
All families	3.7	3.8	3.9	3.9	3.7	3.5	3.3*
PNF households†	–	–	4.0	4.0	3.8	3.5	3.3

SOURCE: Calculated from the census of Canada, various years

* In 1981, datum for "All families" based on families in private dwellings only. Families in collective dwellings are excluded.

† PNF households consist of a census family living alone. See chapter two for more detail. Average size of PNF household is not available prior to 1961.

cally small, families living alone tended to be larger on average than were all families taken together. However, the difference was muted by infrequency of lodging. Further, the difference shrank over time as the number of lodging families declined. As far back as we have good data, the mean sizes of all families and families living alone were quite similar. See Table 20. Between 1961 and 1966, the average household dropped from 3.9 to 3.7 persons, but the average size of a family living alone remained roughly constant.

Why then did mean household size decline during this time? To answer this, one must look at different types of households. The decline in overall average household size can be decomposed into (i) a decline in the average size of each type of household and/or (ii) a shift in the relative number of each type. Let us refer to (i) as a "size" effect and (ii) as a "composition" effect. Table 21 helps us to quantify these two effects using five types of households. First, contrast 1961 with 1966. The average sizes at the two dates are the same for each type of household. In other words, the size effect was zero. Overall average size declined strictly because household composition changed. Small nonfamily households rose from 13 per cent to 16 per cent of all households while the larger multi-family and PNFP (families maintaining a dwelling but with one or more nonfamily individuals present) households became relatively less common. The decline in overall average household size between 1961 and 1966 resulted from a shift away from shared accommodation and toward nonfamily households. All the change between 1961 and 1966 was due to this composition effect.

What happened after 1966? In Table 21, there is evidence of a continuation of the composition effect. Nonfamily units continued to form a larger share of all households, rising to 25 per cent by 1981.[3] Also, the relative numbers of families in various forms of shared accommodation continued to decline. However, after 1966, the average size of each type of household also began to decline. Families living alone shrank in size. So too did households containing families in shared accommodation.

TABLE 21
Average size and percentage distribution (in parentheses) of households by type, Canada, 1961–81

		1961	*1966*	*1971*	*1976*	*1981*
One-family households						
PNF*	Persons/hhld	4.0	4.0	3.8	3.5	3.3
	Percentage	(72)	(72)	(71)	(70)	(67)
PNFP†	Persons/hhld	5.1	5.1	5.0	4.8	4.5
	Percentage	(10)	(9)	(8)	(7)	(6)
Other‡	Persons/hhld	4.5	4.5	4.2	4.0	3.6
	Percentage	(1)	(1)	(1)	(0)	(1)
Multi-family households	Persons/hhld	6.8	6.8	6.4	6.2	6.1
	Percentage	(4)	(3)	(2)	(1)	(1)
Nonfamily households	Persons/hhld	1.4	1.4	1.4	1.3	1.2
	Percentage	(13)	(16)	(18)	(21)	(25)
All households	Persons/hhld	3.9	3.7	3.5	3.1	2.9
	Percentage§	(100)	(100)	(100)	(100)	(100)

SOURCE: Computed from published census reports, various years
* The concept of a "primary" family changed in 1981. Previously, a family was defined to be primary if the household head was a member. In 1981, the notion of head was replaced by maintainer. See chapter two for details. However, the household maintainer need not reside in the household, e.g. where a female-lone-parent family is maintained by a separated husband. In such cases, the 1981 Census terms this a secondary family (even if living alone). Thus, some households that would have been classified as PNF units prior to 1981 were counted as "Other" one-family units in 1981.
† Consists of families maintaining a dwelling with additional persons present who do not form a second family. Households with two or more families present are included in "Multi-family households."
‡ Includes any household with one family present but where the household head/maintainer is not a family member. See asterisked note.
§ Column percentages may not add due to rounding.

And the average size of a nonfamily household fell from 1.4 to 1.2 persons, reflecting the emergence of the one-person household. Thus, in the decline of overall average size after 1966, there were both composition and size effects.

Several conclusions can be drawn from this discussion. The decline in average size can usefully be described in terms of composition and size effects. The decline in household size between 1961 and 1966 was attributable strictly to composition effects. After 1966, size effects also became significant.

THE AGE COMPOSITION OF HOUSEHOLDS

It is misleading to look at average household size without considering changes in age composition. The coming into adulthood of the baby

TABLE 22
Household heads (000s) by age, sex, and marital status, Canada, 1961-81

Age group	*Marital status*	*Households*			
		1961	*1966*	*1976*	*1981*
(a) Men					
15-24	Total	159	227	444	427
	MSPM*	145	192	323	302
	Other†	14	35	121	-
25-34	Total	895	955	1,482	1,600
	MSPM	862	901	1,309	1,429
	Other	33	54	173	-
35-44	Total	1,001	1,103	1,197	1,309
	MSPM	970	1,045	1,101	1,221
	Other	31	58	96	-
45-54	Total	833	928	1,127	1,115
	MSPM	791	860	1,021	1,023
	Other	42	68	106	-
55-64	Total	558	688	851	909
	MSPM	507	582	753	823
	Other	51	73	98	-
65 or older	Total	516	562	728	812
	MSPM	415	438	582	682
	Other	101	124	146	-
(b) Women					
15-24		20	42	140	245
25-34		44	60	197	438
35-44		71	87	143	282
45-54		104	124	179	256
55-64		123	148	228	305
65 or older		231	288	450	583

SOURCE: Computed from published census reports, various years
* 'MSPM' is a married male household head in 1961, a married (spouse-present) male household head in 1966 and 1976, and a husband in a husband-wife primary family in 1981. In 1981, a family is not primary is the household maintainer lives elsewhere. In 1966 and 1976, the same family would have been a primary unit if the head was a family member.
† 'Other' is calculated by subtracting MSPM from total.

boomers, in the late 1960s and the 1970s, for example, had several effects on household size. For one, the size of the parental household shrank as the baby boomers began to leave home. As young adults, they sought out their own living arrangements. Some lived alone. Others married and as young, childless marrieds also contributed to a reduction in average family size. These effects were exacerbated by the tendency to delay or forgo child birth and marriage. These effects are consistent with the decline in household size and the rise in living alone after 1966. To what extent, therefore, are these household size and composition changes simply manifestations of changing age composition?

We can derive rough estimates of the effect of the aging baby boom on household formation from published census material. Prior to 1961, Canadian censuses did not detail household heads by age, sex, and marital status. Further, as noted in chapter two, the notion of headship was abandoned in the 1981 Census in favour of household maintainer. As well, in 1981, a greater percentage of wives were reported as maintainers, making historical comparisons by sex more difficult. The following discussion is, therefore, constrained largely to the period between 1961 and 1981. Household counts for this period are presented in Table 22. Between 1961 and 1976, the greatest increases in household formation occurred among heads aged under 35. As this period marked the start of the marriage bust, the number of households headed by a husband-wife couple increased only moderately. Household formation was more rapid among not-marrieds. Note also the rapid formation among older women. Many of these were widows living alone. With women increasingly outliving their husbands, the number of widows becoming household heads climbed sharply.

Households with Married (Spouse Present) Male Head

What role did such changes in age composition and marital status play in the decline in average household size? To answer this question, we have to disaggregate the household size data further. Consider the set of households (denoted here as MSPM) headed by a married male with spouse present. Average household size data are presented in Table 23. The average size of an MSPM household fell from 4.2 persons in 1966 to 3.7 persons in 1976, paralleling the declining size of families living alone. The remainder of Table 23 disaggregates these average sizes by the age of the husband. Underneath each average size, in parentheses, is the percentage of all MSPM households headed by a husband of that age.

The latter figures show a shift attributable to the aging baby boomers. Between 1966 and 1976, the ranks of the under-35s swelled. For example, 25-34-year-old heads rose from 22 per cent to 26 per cent of all MSPM households. However, this was mirrored by a relative drop in 35-44-year-old heads, a consequence of low fertility in the 1930s. These two changes were offsetting, and the net effect on average household size was negligible. To show this, suppose the average size for each age group had not fallen from 1966 to 1976. Considering the age composition change alone, average MSPM household size in 1976 would have been

$$0.06(3.0) + 0.26(4.3) + 0.22(5.2) + 0.20(4.6) + 0.15(3.5) + 0.11(2.7) = 4.2,$$

TABLE 23
Average size and percentage distribution (in parentheses) of households headed by a married (spouse-present) male by age group, Canada, 1966 and 1976

Year*		*Age of husband*						*All*
		15–24	*25–34*	*35–44*	*45–54*	*55–64*	*65+*	MSPM
1976	Persons/hhld	2.6	3.5	4.6	4.3	3.1	2.4	3.7
	Percentage	(6)	(26)	(22)	(20)	(15)	(11)	(100)†
1966	Persons/hhld	3.0	4.3	5.2	4.6	3.5	2.7	4.2
	Percentage	(5)	(22)	(26)	(21)	(14)	(11)	(100)

SOURCE: Computed from published census reports, 1966 and 1976
* In 1966, married women could not head a household if spouse was present. In 1976, either spouse could be head. The 1976 data exclude households with a married (spouse-present) female head as the corresponding age of husband could not be calculated from published census reports. The 1976 average sizes overestimate the average sizes of all husband-wife households because those where the wife was head were typically smaller. However, the bias is small, since relatively few wives were designated heads.
† Row percentages may not add due to rounding.

the same as in 1966. The net effect of the change in age composition, the primary feature of which was the aging of the baby boom, was negligible. Rather, it was the decline in average size in each age group that produced the decline in overall average MSPM size.

Other Households

Next, consider households headed by someone who is not living with a spouse. Such living arrangements can take several forms, from lone-parent households to nonfamily households. However, the dominant living arrangement here is the person living alone. The rapidly increasing number of one-person households has already been noted. In this section, we focus on the incidence of living alone by age and sex.

Data availability limits our analysis to the period from 1966 to 1976. Relevant data are presented in Table 24. The patterns are similar for both sexes. In 1966, 59 per cent of all male household heads without a spouse lived alone. Between 1966 and 1976, the proportion rose: from 59 per cent to 65 per cent for men and from 46 per cent to 54 per cent for women. Again, we see the baby boomer influence. Male heads aged under 35 accounted for 21 per cent of all such households in 1966 (14 per cent for women) and 39 per cent in 1976 (25 per cent for women). At the same time, the net effect of changing age composition was small. This is because the baby boom swelled the ranks of the 15–24-year-old heads, who were among the least likely to live alone, as well as the ranks of the 25–34-year-olds, who were much more likely to do so. A calculation sim-

TABLE 24
Percentage in one-person households and percentage distribution (in parentheses) of households with head not living with spouse,* by age group, Canada, 1966 and 1976

Year	*Age of household head*						*All*
	15–24	*25–34*	*35–44*	*45–54*	*55–64*	*65+*	*ages*
(a) Male heads							
1976	58	67	63	62	66	71	65
	(16)	(23)	(13)	(14)	(13)	(20)	(100)†
1966	55	65	62	56	58	60	59
	(8)	(13)	(14)	(17)	(18)	(30)	(100)
(b) Female heads							
1976	53	44	28	38	59	72	54
	(10)	(15)	(11)	(13)	(17)	(34)	(100)
1966	42	38	32	35	48	56	46
	(6)	(8)	(12)	(17)	(20)	(38)	(100)

SOURCE: Computed from published reports of the 1966 and 1976 censuses of Canada
* "Not living with spouse" includes single, widowed, divorced, or separated persons.
† Row percentages (in parentheses) may not add due to rounding.

ilar to that used above shows that if the proportion of heads living alone at each age had not changed between 1966 and 1976, the overall percentage living alone in 1976, among male heads, would have been

$$0.16(55) + 0.23(65) + 0.13(62) + 0.14(56) + 0.13(58) + 0.20(60) = 59,$$

about the same as in 1966. For women, the corresponding calculation yields an incidence of 45 per cent; again similar to the 1966 level. In other words, the rising overall incidence of one-person households had little to do with changing age composition. Instead, it was primarily the result of a rising incidence of living alone within each age group.

These results suggest that a changing age composition and, in particular, the aging baby boom did not have a large effect, between 1966 and 1976, on either the average size of a household or on the incidence of living alone. Changes in age structure did have some effects, but these were in opposite directions, with little net effect on either average household size or the incidence of persons living alone. However, we are interpreting "age structure" in a special way here. We have used the figures in parentheses in Tables 23 and 24 which are the percentage distributions of households by age of head. This is not the same thing as looking at the percentage distribution of population by age, because the propensity to be a household head changed over time. To understand more clearly the role of population change in household formation, we need to introduce another concept – the headship rate.

TOTAL HEADSHIP RATES

A headship rate is the ratio of the number of households headed by a person of a given type divided by the total number of persons in the population of that type. As a percentage, a headship rate cannot exceed 100. As an example, suppose we want to calculate a total headship rate (THR) for widowed men aged 55-64 in 1976. Of the 32,395 such men in Canada in 1976, 25,745 were heads of household, for a THR of 80 per cent. This is a "total" headship rate because it includes all kinds of households. In contrast, we might be interested instead in the number of one-person households headed by widowed men aged 55-64. There were 13,540 such households in 1976, for a one-person headship rate (1PHR) of 42 per cent.

What about the remaining persons in the group? If not heads of household, what were they? Table 25 provides some detail for 55-64-year-old widowers. Of those who did not head any household, 2,820 were related to the head of household (though not in a husband-wife or parent-child family), and 1,915 were lodgers, employees, or partners to a head. Some lived in collective dwellings, and a small number headed a secondary lone-parent family (either living with relatives or lodging).[4]

A headship rate is simply the propensity for individuals in a given group to be a household head. It is helpful when one looks at the relationship between demographic change and household formation. When, for example, a headship rate rises over time, it implies an increase in the number of households for a given size and composition of population.

The usefulness of a headship rate is based, however, on an important condition: the stability of the definition of head. As discussed in chapter two, there have been different ways of defining household headship. The definition used in censuses from 1951 to 1971 remained the same. However, in 1976, the definition was expanded to include either spouse, whereas earlier a wife could not be head if her husband was present. As might be expected, in 1976 the headship rates for married women rose and the rates for married men either fell or increased only marginally. Part of this change was induced simply by the change in definition and did not indicate a change in the pattern of household formation. Such definitional changes can make interpretation of changes in headship rate problematic.[5]

Let us now look at how total headship rates changed over the postwar period. Prior to 1961, published census data do not disaggregate headship by marital status, so let us begin with headship rates simply by age group and sex. These are shown in Table 26. Note the continued upward trend in all these total headship rates through 1976. After 1976, the decline in male headship reflects the increasing incidence of wives as household maintainers.

TABLE 25
Household and family status of widowed men aged 55-64, Canada, 1976

	Population	*Heads of household*
In private households		
In families		
Primary lone parents	9,240	9,240
Secondary lone parents	320	
Not in families		
Household heads		
Living alone	13,540	13,540
Others present*	2,965	2,965
Related to head	2,820	
Lodgers	1,635	
Employees	50	
Partners	230	
In collective households, other	1,595	
Total	32,395	25,745

SOURCE: Computed from published reports of the 1976 Census of Canada
* "Other" includes individuals in private dwellings whose relationship to head of household could not be established.

Are these postwar trends merely a continuation of a process that began much earlier? It is difficult to make comparisons with the prewar period in Canada because of the difference in household definition. However, two U.S. studies suggest that the postwar rise in headship rates was unprecedented. One of them, Kobrin (1976b), finds substantial growth after 1950 but little change in household headship rates between 1940 and 1950. The other, Hickman (1974), used 1940 headship rates to estimate net household formation each year between 1921 and 1940. Hickman then compared the estimate with the actual net number of households formed in each year. Where the two figures are the same, it is consistent with the argument that the household headship rates for that year are the same as those in 1940. Hickman found that the two figures were similar for each year during the 1920s; the estimate considerably exceeded actual counts during the early 1930s and then fell below them in the later 1930s. Thus headship rates appear to have been relatively constant during the 1920s, but fell sharply during the Depression, readjusting thereafter. In this sense, the continued rise of headship rates in the postwar period, far beyond anything that could be considered a recovery from the Depression, is unprecedented.

Let us now consider some of the variations in Canadian postwar headship rates evident in Table 26. Traditionally, these rates were highest for men aged 35-64, most of these men being married and maintaining a family household. Total headship rates were lower – but increased

TABLE 26
Total headship rates (%) by age group and sex, Canada, 1956-81

Age group	*1956**	*1961*	*1966*	*1971*	*1976*	*1981†*
(a) Men						
15-24		12	14	17	19	18
25-34	39	71	76	79	80	76
35-44	82	84	86	89	90	88
45-54	84	87	89	90	91	89
55-64	84	85	88	89	91	88
65 or older	75	77	78	80	82	80
(b) Women						
15-24		2	3	4	7	11
25-34	2	4	5	8	12	21
35-44	5	6	7	9	12	19
45-54	10	11	12	13	15	20
55-64	18	19	20	22	24	27
65 or older	29	32	35	37	41	43

SOURCE: Adopted from Wargon (1979a, 63) and published reports of the 1976 and 1981 censuses of Canada

* In 1956, youngest age group available is 15-34. Data shown in row "25-34" are for this youngest age group.

† In 1981, headship is based on household maintainer, or person 1 if maintainer was not coresident. In 1976, it became possible for a wife to be identified as head.

quickly over time - among the under-35s and to a lesser extent among the elderly. Total headship rates for women were considerably lower, in large part because of the definition of headship used prior to 1976. Even in 1976, when a wife could be head, few chose to do so. With the introduction of the household maintainer concept in 1981, this became more commonplace. As a result, total headship rates were low for women, except among the elderly, where widowhood and headship often went hand in hand. In spite of the definitional bias against women, their headship rates increased quickly during the postwar period. In part, this reflects changes in marital status. The rising incidence of divorce and bachelorhood during the marriage bust accounts for some of the increase in headship rates among both women and men. To assess the effect of this, let us next disaggregate headship rates by marital status.

Married Persons

In 1966, there were 228,967 married men aged 15-24. Of these, 84 per cent headed a household in which the spouse was also present. An additional 1 per cent headed a household where the spouse was absent. There were 475,614 married women of the same age. Remember that in the

TABLE 27
Spouse-present and spouse-absent headship rates (%) for married men and women, Canada, 1966–81

Household type, sex, year	*15–24*	*25–34*	*35–44*	*45–54*	*55–64*	*65+*
Spouse-present household						
Married* male head						
1981	78	87		90		85
1976	85	91	94	94	93	89
1966	84	92	94	94	93	88
Married female head						
1981	9	7		5		8
1976	2	2	1	1	1	2
1966	n.a.	n.a.	n.a.	n.a.	n.a.	n.a.
Spouse-absent household						
Married male head						
1981	1	2		3		2
1976	2	2	2	3	3	3
1966	1	1	1	2	3	3
Married female head						
1981	2	4		4		3
1976	2	3	4	4	4	4
1966	1	2	2	3	3	3

SOURCE: Computed from published reports of the 1966, 1976, and 1981 censuses of Canada

* The 1981 Census distinguishes between married and separated household maintainers. A separated maintainer has no spouse present. In this table, it is assumed that all married maintainers have a spouse present. In 1966, a married woman could not head a household if her spouse was present.

1966 Census, only those married women whose husband was absent could be a household head. There were 4,473 such household heads, for a headship rate of just under 1 per cent. Headship rates for other age groups of marrieds between 1966 and 1981 are shown in Table 27.

The spouse present headship rates (SPHR) for married men dropped in 1976 and again in 1981. This is likely attributable to the changing definition of headship. The small 1976 SPHR for married women indicates that few were designated head even though the Census definition had been liberalized. With the advent of the household maintainer concept, increases in female SPHR were more substantial.

The spouse absent headship rates (SAHR) display more variation. There were substantial increases in SAHR between 1966 and 1981 in the younger age groups. Older age groups were characterized by smaller increases, even decreases. Two factors were at work here. First, there was a slight rise in the incidence of marital separation after 1966, mainly among the under-55s. Second, these separated persons became increasingly likely to head a household. Both contributed to the rise in SAHR among the younger groups.

Widowed and Other Marital Statuses

In 1966 Census reports, not-married heads are disaggregated only by whether or not they were widowed; i.e. singles are lumped together with divorced persons as "other" marital statuses. The 1976 and 1981 censuses provide separate data for singles and divorced persons. Total headship rates for widows and widowers are presented in Table 28. Since widowhood was uncommon among young adults, these rates are shown only for older age groups. Note the substantial differences between men and women, especially under age 45, widows being more likely to head a household than were widowers. Note also the substantial increases in total headship rates over time. The story was similar among other not-marrieds. Headship was more common among younger women than younger men, and headship rates grew markedly between 1966 and 1981. Also, note that among other not-marrieds, headship rates were low among the young, reflecting the prevalence of living in a parental home.

Bernard (1975), in comparing the living arrangements of U.S. divorced men and women in 1970 and 1974, finds that divorced women were more likely than men to be household heads. Bernard attributes this to the presence of children. Ordinarily, women obtained custody of children upon divorce, and subsequent living arrangements reflected this. Since he typically did not have custody of children, a divorced man could more easily live as a lodger or otherwise share space and hence not be a head of household. With children present, such shared space options were less attractive or available to divorced women; they were more likely to maintain a dwelling of their own. Also, divorced men who did retain custody of children were much more likely to move in with someone else, perhaps parents, who could assist in child rearing. Again, this could contribute to a lower headship rate among divorced men.

These marital-specific headship rates provide valuable additional information on the overall pattern of postwar household formation. Chapter three examined postwar patterns of population growth, changing age composition, and changing marital statuses. These undoubtedly contributed to the pace of household formation. The marital-specific headship rate data above suggest, in addition, that the living arrangements of particular kinds of individuals changed. In other words, even if there had been no baby boom, marriage rush, or marriage bust, the pattern of household formation probably still would have been altered. For various reasons, families and nonfamily individuals sought new living arrangements, typically in smaller households than before. This pattern

TABLE 28
Total headship rates (%) for widowed and other not-married men and women, Canada, 1966–81

<table>
<tr><th>Marital status, sex, year</th><th>15–24*</th><th>25–34</th><th>35–44</th><th>45–54</th><th>55–64</th><th>65+</th></tr>
<tr><td>Widowed</td><td></td><td></td><td></td><td></td><td></td><td></td></tr>
<tr><td>Men</td><td></td><td></td><td></td><td></td><td></td><td></td></tr>
<tr><td>1981</td><td>–</td><td colspan="2">78</td><td colspan="2">83</td><td>63</td></tr>
<tr><td>1976</td><td>–</td><td>61</td><td>79</td><td>83</td><td>80</td><td>60</td></tr>
<tr><td>1966</td><td>–</td><td>46</td><td>69</td><td>75</td><td>64</td><td>53</td></tr>
<tr><td>Women</td><td></td><td></td><td></td><td></td><td></td><td></td></tr>
<tr><td>1981</td><td>–</td><td colspan="2">89</td><td colspan="2">86</td><td>67</td></tr>
<tr><td>1976</td><td>–</td><td>82</td><td>91</td><td>98</td><td>84</td><td>66</td></tr>
<tr><td>1966</td><td>–</td><td>77</td><td>85</td><td>81</td><td>72</td><td>59</td></tr>
<tr><td>Other not married†</td><td></td><td></td><td></td><td></td><td></td><td></td></tr>
<tr><td>Men</td><td></td><td></td><td></td><td></td><td></td><td></td></tr>
<tr><td>1981</td><td>8</td><td colspan="2">46</td><td colspan="2">61</td><td>56</td></tr>
<tr><td>1976</td><td>6</td><td>34</td><td>43</td><td>48</td><td>53</td><td>50</td></tr>
<tr><td>1966</td><td>2</td><td>17</td><td>25</td><td>34</td><td>43</td><td>43</td></tr>
<tr><td>Women</td><td></td><td></td><td></td><td></td><td></td><td></td></tr>
<tr><td>1981</td><td>10</td><td colspan="2">62</td><td colspan="2">68</td><td>54</td></tr>
<tr><td>1976</td><td>8</td><td>48</td><td>59</td><td>56</td><td>53</td><td>48</td></tr>
<tr><td>1966</td><td>3</td><td>23</td><td>30</td><td>35</td><td>38</td><td>38</td></tr>
</table>

SOURCE: Computed from published reports of the 1966, 1976, and 1981 censuses of Canada

* There were too few widowed persons aged 15–24 to estimate headship rates reliably.

† Single and divorced persons

was especially marked, during the period from 1966 to 1981, for persons living without a spouse. Below, "undoubling" of family units is also shown to have been important, at least up to about 1966.

In spite of large increases between 1966 and 1981, headship rates for not-marrieds were still low compared to their U.S. counterparts. Sweet (1984, 134) reports, for example, that about 84 per cent of U.S. widows over 64 were household heads in 1980, up from 74 per cent in 1970. In Canada, only 66 per cent were heads at mid decade. Similar differences are found at other ages and marital statuses.

ONE-PERSON HEADSHIP RATES

The concept of a rising total headship rate is difficult to grasp. We know that it means that members of a group have become more likely to head a household. However, it does not tell us much about the composition of that household. In this regard, a one-person headship rate is more informative. It tells us the percentage of a population group that is living alone. We have already noted that the rise in persons living alone was a

TABLE 29
Percentage distribution of persons living alone by age and marital status, Canada, 1976

Age group	*Single*	*Married**	*Widowed*	*Divorced*	*Total*
15-24	11	1	0	0	12
25-34	13	2	0	2	17
35-44	5	2	0	2	8
45-54	5	2	2	2	11
55-64	5	2	8	2	16
65 or older	6	2	27	1	36
Total	44	12	37	8	100

SOURCE: Computed from published reports of the 1976 Census of Canada
* Includes persons who are separated (i.e. spouse absent)

dominant aspect of postwar household formation. To understand why average household size declined, it is important to know why the number of one-person households increased so rapidly.

When thinking of persons living alone, a common image is of the young single adult. In the real estate industry, for example, the smaller units geared to such consumers are referred to as "bachelor" apartments, "studios," "junior one bedrooms," or "singles" housing. However, although young singles contributed substantially to the growth in living alone, the living arrangement was dominated by the elderly. Even in 1976, when the baby boom's impact on the number of young adults was peaking, fewer than 3 in 10 persons living alone were under 35; just over 1 in 10 were under 25. As seen in Table 29, half of all persons living alone in Canada were aged 55 or older, many widowed. In the United States, the age distribution was even more skewed; Glick and Norton (1977, 30) report that, in 1976, 43 per cent of those living alone were 65 or older, and only 23 per cent were under 35. In both countries, the age distribution of one-person households was bimodal, dominated by elderly widows and, to a lesser (but growing) extent, young singles.

While changes in age and marital composition accounted for some of the increase in living alone, much was directly attributable to increasing one-person headship rates (1PHR). Detailed census data on living alone have been available only since 1971. As displayed in Table 30, there were substantial increases in these 1PHR between 1971 and 1981. In other words, nonfamily individuals became much more likely to live alone. Michael et al (1980, 40) present 1PHR for 1950 and 1976, which show the same upward trend in the United States. The U.S. 1PHR in 1976 are comparable to those for Canada. Kobrin (1976b, 237) presents additional data for the United States back to 1940, which suggests that the increasing tendency to live alone there may not have begun until the 1950s.

TABLE 30
One-person headship rates (%) by age, sex, and marital status, Canada, 1971 and 1981

Age group	*Year*	*Single*	*Married**	*Widowed*	*Divorced*
(a) Men					
15-24	1971	2	0	4	10
	1981	5	1	13	22
25-34	1971	15	1	10	22
	1981	30	2	27	42
35-44	1971	21	1	12	27
	1981	36	2	19	44
45-54	1971	26	1	19	32
	1981	44	2	27	45
55-64	1971	31	2	34	41
	1981	42	2	45	54
65 +	1971	32	2	35	44
	1981	40	2	47	58
(b) Women					
15-24	1971	2	0	7	8
	1981	6	1	12	16
25-34	1971	18	0	6	11
	1981	35	1	12	21
35-44	1971	22	0	7	14
	1981	37	1	9	18
45-54	1971	25	1	19	27
	1981	36	1	24	32
55-64	1971	28	2	38	43
	1981	40	2	49	56
65 +	1971	28	2	41	46
	1981	39	3	53	66

SOURCE: Computed from Harrison (1981, 30-1) and the 1981 Census of Canada
* Includes persons who are separated (i.e. spouse absent)

There were substantial age differentials in 1PHR. Among singles, young adults were the least likely to live alone. Only 5 per cent of 15-24s, and 30 per cent of 25-34s, lived alone in 1981, compared with 40 per cent of those 55 or older. In chapter six, it is argued that this age differential reflects differences in incomes or housing costs. Typically, living alone was expensive. Without roommates to help spread the cost of housing, it was necessary to have a good income. The principal exceptions to this were the elderly, who, during the 1960s and 1970s, benefited from extensive construction of senior citizen housing. Such housing was typically subsidized by taxpayers and made living alone affordable to a broad group of individuals with modest incomes.

Typically, 1PHR were low also for young ever-marrieds. This was especially true of married and widowed persons in general and of divorced women in particular. This reflects, in part, the presence of young children or a spouse. Younger divorced men, however, were more likely to live alone. Some of the largest relative increases in 1PHR in the 1970s were among younger ever-marrieds. This could be a consequence of declining fertility. As couples increasingly delayed or forwent child birth, the probability that they would be childless at separation, divorce, or widowhood rose correspondingly – and hence the likelihood of living alone.

If marital dissolution occurred while the partners were young, it was not uncommon for at least one to move back to a parental home. Glick (1976, 328–9) notes that, in the United States in 1970, about 40 per cent of separated or divorced persons aged 18–24 lived with their parents. This incidence declined with age, and the incidence of living alone rose correspondingly. The availability of this alternative living arrangement helps to explain why 1PHR are lower among younger separated or divorced persons.

The decision to live alone at other ages is also partly constrained by available alternatives. For elderly widows, in particular, an alternative to living alone is to live with an adult child. Sometimes, this alternative is not feasible – the child may be too far away, or the parent and child (or child's family) may be incompatible. However, other things being equal, the more children borne, the more likely it is that a widow would have at least one viable alternative to living alone. Kobrin (1976a) uses this argument to explain the rising incidence of living alone among widows in the 1960s and early 1970s. Remember that these women were from the cohort whose principal child bearing years were during the 1930s. Having had few children, they were less likely later to live with one of them. Thus, the increasing number of postwar widows living alone may partly have resulted from the prewar fertility slump.[6] Supporting evidence is found in Harrison (1981, 41–50). After controlling for the widow's income and number of children ever borne, he finds that Canadian widows who had borne at least three children were less likely to be living alone than were other widows at the same income and age. In a similar U.S. study, Chevan and Korson (1972) found that childless widows were most likely to live alone and that the probability of living alone declined the more children ever borne by the widow.

Harrison's analysis, by controlling for income, raises a broad question. Just how important were rising incomes in the formation of one-person households? On the one hand, these occurred apace. The postwar period was prosperous. The incomes of most categories of individuals and families rose rapidly at the same time as increasing numbers of indivduals were choosing to live alone. Since living alone is expensive, a causal relationship is plausible. However, Beresford and Rivlin (1966, 254), among others, caution that real income growth was not the only factor at work. They point out that real incomes also rose in the first half of the century, without a corresponding increase in the number of one-person households. In their view (40): "To assert that separate living has been associated with higher incomes since 1940, one must also assert that a basic shift in tastes occurred at about that time after which people tended to use their rising incomes to purchase additional privacy." This view does not, however, address the questions of why tastes changed and why privacy became more important.

Living alone can be seen as a choice from among alternatives. For the nonfamily individual, the other alternatives have traditionally been to live with one's family or relatives, in a collective dwelling, as a lodger in a small rooming-house or family home, or in a dwelling with one or more partners. Over the postwar period, the living alone option became more widely available, relatively less expensive, or more attractive than these alternatives.

In part, this reflected the shrinking of some alternatives. On the one hand, there was a concerted effort by postwar governments across Canada to reduce the size of the institutionalized population. From psychiatric hospitals to nursing homes, the collective dwelling stock was seen as an inefficient way to provide housing and support services. Efforts were made to rehouse residents within smaller group homes and private housing. In addition, support services such as "meals on wheels" and visiting nurses were developed to enable potential inmates to remain in their own homes. As a result, postwar counts of the collective populatiom stagnated, even as the number of nonfamily individuals increased dramatically. Also important was the decline in commercial lodging among family households, a consideration explored in the next section of this chapter. Another factor was the proliferation of local building code and zoning restrictions that made it increasingly difficult to supply lodging legally.

In part, living alone became more popular because it became more affordable. In chapter eight, evidence is presented to suggest that the cost of rental housing in particular increased less quickly than did the prices of most other goods and services. As rental housing became more affordable, it encouraged more individuals, particularly those of modest income, to live alone. What brought about this improved affordability is not clear. I think it reflects in part improved efficiency in the construction of housing in general, and high-rise housing in particular. I think it had something to do also with income tax incentives that resulted in an "oversupply" of new rental housing. However, a discussion of factors shaping the supply of housing is beyond the scope of a book that is fundamentally concerned with demand.

Living alone became more popular in part because it became more feasible. Maintaining one's own household has traditionally involved a certain amount of labour. Activities such as housecleaning, repair and maintenance, laundry, and food preparation require time, effort, and physical dexterity. Over the postwar period, such activities were made somewhat less onerous or demanding for persons living alone. We have already discussed the improvements in home-making technology that made some of these chores more manageable and the development of publicly funded support services that provided home-making services for the elderly and disabled. In addition, a variety of private market goods and services arose to fill the special needs of persons living alone, e.g. emergency hailers, commercial maid services, and concierge-equipped condominiums.

Some of the increase in living alone reflected fairly modest changes to the existing housing stock. A family of four with a lodger living downstairs might choose to wall off the basement area, provide a separate entrance from the outside, and construct separate bathroom and kitchen areas. In so doing, it creates two dwellings (the lodger's basement flat and the family's upstairs unit) where previously there had been only one. Concurrently, the five-person household is split into two: a four-person household and a person living alone. Undoubtedly, a portion of the postwar increase in living alone was due to such structural redefinitions.

LIVING ARRANGEMENTS OF FAMILIES

Let us now turn our attention back to the living arrangements of families. To what extent was the postwar rise in total headship rates attributable to "undoubling" in its various forms? Published census data indicate that undoubling was important, although in different forms in the early and later postwar years. See Table 31. In 1951, about 90 per cent of families maintained their own dwellings, the remainder lodging with one or more other families or individuals. From 1951 to 1966, this proportion rose steadily, to 96 per cent, and it remained constant thereafter. The declining number of lodging families was one form of undoubling, but it apparently was significant only up to about 1966.

Just how important was the postwar undoubling of families in spurring household formation? A simple calculation is illustrative. In 1951, 90 per cent of families maintained a dwelling. If this percentage had not changed through 1981, there would have been only (0.90 × 6,325,000) = 5.7 million such families, 0.4 million fewer than actually observed. The number of families maintaining a dwelling increased by 3.1 million between 1951 and 1981; undoubling thus accounts for about 13 per cent of the increase. Thus undoubling in the sense of fewer families lodging

TABLE 31
Families (000s) by living arrangement, Canada, 1951–81

Year	Total	*Maintaining own dwelling** Total	*Living alone*	*Other present*	*Not maintaining own dwelling*†
1951	3,287	2,967	n.a.	n.a.	321
1956	3,712	3,426	n.a.	n.a.	286
1961	4,147	3,912	3,263	649	235
1966	4,526	4,346	3,755	591	180
1971	5,076	4,915	4,286	629	161
1976	5,808	5,605	5,026	579	203
1981	6,325	6,133	5,556	577	192

SOURCE: Drawn from published census reports, various years
* Prior to 1981, a family maintains a dwelling if the household head is a family member. In 1981, a family is deemed to maintain a dwelling if the household maintainer is a family member.
† In censuses prior to 1981, families "not maintaining own dwelling" are lodging with another person or family. In 1981, a family was also deemed not to maintain a dwelling (even if it lived alone) if the household maintainer lived elsewhere.

made an important contribution to the net formation of family households.

Another form of undoubling is the spinning off of nonfamily members from family households. In 1961, 79 per cent of all families lived alone. By 1981, this had risen to 88 per cent. The increasing incidence of families living alone accounted for about one-fourth of the total net formation of PNF households between 1961 and 1981. The number of PNFP households (i.e. families maintaining a dwelling but with other persons present) remained roughly constant during this period but fell rapidly as a percentage of all families.

Chapter three notes several postwar changes in family formation. The effects of changing nuptiality and fertility on family formation and on the growth of lone parenthood were extensive. Also significant were changes in age distribution. What effects did these changes have on living arrangements? Did the rise of lone parenthood, for example, change the likelihood that a family would live alone? Did living arrangements vary with age, and if so did the aging baby boomers have an impact on aggregate counts such as in Table 31?

Unfortunately, there are few suitably disaggregated historical data available concerning the living arrangements of families. However, since 1951, censuses have reported the number of families maintaining a dwelling by type of family and age of head. Table 32, computed from these data, shows the proportion of all families by type (i.e. husband-wife, male lone parent, or female lone parent) and age (of husband or lone

TABLE 32
Percentage of families maintaining own dwelling by family type and age of husband or lone parent, Canada, 1951-81

Age	*1951**	*1956*	*1961*	*1966*	*1971*	*1976*	*1981†*
(a) Husband-wife families							
15-24			82	89	91	96	94
25-34	83	87	93	95	97	98	97
35-44	94	95	97	98	99	99	98
45-54	96	97	98	99	99	99	99
55-64	97	97	98	98	99	99	98
65 or older	93	94	95	96	96	97	97
(b) Male lone-parent families							
15-24			32	17	70	68	54
25-34	44	56	58	58	86	77	77
35-44	70	74	77	81	90	90	91
45-54	84	85	86	91	95	95	94
55-64	89	90	91	93	95	96	95
65 or older	89	90	92	93	94	94	93
(c) Female lone-parent families							
15-24			28	42	62	79	82
25-34	43	44	59	70	82	88	88
35-44	72	75	80	86	90	94	93
45-54	86	86	88	92	94	96	94
55-64	89	90	91	93	95	96	94
65 or older	89	91	93	94	95	96	95

SOURCE: Computed from published census reports, various years
* In 1951 and 1956, "Under 35" is the youngest age group provided. These data are shown above in the rows marked "25-34."
† 1981 data are defined differently from earlier years. Prior to 1981, all families living alone, for example, were defined to be "maintaining own dwelling." However, in 1981, families living alone but wherein the household maintainer resided elsewhere were defined to be "not maintaining own dwelling."

parent) maintaining a dwelling. In making historical comparisons, remember that the 1981 data are based on household maintainer, rather than household head. Prior to 1981, any family living alone, for example, was deemed to be maintaining a dwelling. However, in 1981, families living alone but with the household maintainer residing elsewhere (e.g. a separated wife and children maintained by a nonresident husband) are defined to be not maintaining a dwelling. This accounts for some of the drops between 1976 and 1981 observable in Table 32.

Consider first the data for husband-wife families. Note the consistently rising proportions from 1951 to 1976. Across all age groups, husband-wife families became more likely to maintain a dwelling. Families in the middle age groups were always unlikely to lodge. The sharpest declines in lodging were among young (and to a lesser extent elderly) couples. The small reversals in 1981 could be a result simply of definitional change. Beresford and Rivlin (1966) present comparable data for the United States between 1940 and 1960. They also find substantial increases in the proportion of married couples maintaining their own dwelling after 1950. However, they found little change between 1940 and 1950 in the level of undoubling. This suggests that the rise of the family living alone was a postwar phenomenon.

Panel (b) presents similar data for lone-parent families headed by men. As with husband-wife units, between 1951 and 1976 lone-parent families became more likely to maintain a dwelling. Lone-parent families were initially more likely than their husband-wife counterparts to lodge, but the difference diminished considerably in the ensuing decades. By 1981, the largest remaining differences were among the under-35s; presumably this reflects the lodging of young male lone parents with relatives such as grandparents who could assist in child rearing.

Similar information for female lone parents is presented in panel (c). The living arrangements of these families changed over time in a similar fashion to those for male lone parents. However, note the rate of growth among under-25s in the proportion maintaining a dwelling. In 1961, only 28 per cent of female lone parents aged 15-24 maintained a dwelling: compared to 32 per cent of men. By 1981, the proportions had risen to 82 per cent for women, compared to 54 per cent for men. Beresford and Rivlin (1966) note a similar change in the United States. An important difference in living arrangements opened up between men and women in this regard.

Undoubling, in the sense of families choosing to live alone, was another important aspect of household formation in postwar Canada. The data in Table 31 suggest that undoubling has been under way since at least 1951. When did this trend first emerge? What caused it?

It is instructive to think of lodging in terms of factors shaping the demand for lodging and the decisions of households to supply lodging. Modell and Hareven (1973) suggest that the decline of lodging in the United States began only after 1920. From at least 1850 up to 1920, they argue (469) that about 15 per cent to 20 per cent of American families had lodgers living with them. By 1930, this had fallen to 11 per cent, and by 1970, under 5 per cent. Modell and Hareven argue that lodging used to be commonplace for several reasons. First, separate housing (especially in urban areas) was in short supply because of rapid population growth. Persons, or families, who might have wanted their own housing could not obtain (or afford) it and had to lodge. Second, lodging was often with nonrelatives, and, for young adults, this meant a desirable degree of independence from family. In the absence of separate accommodation, a concern for privacy and independence made lodging preferable to living in an extended family household. These were factors underlying the demand for lodging. Third, taking in lodgers meant a stable source of income for the family. The layoff, disability, or death of a breadwinner was often a severe financial blow to the family, the effect of which could be ameliorated by a lodger's rent. Finally, taking in lodgers was a way of gainfully employing women in a family at a time when women typically did not work outside the home. These latter two were factors underlying the supply of lodging accommodation.

Modell and Hareven argue that lodging declined for several reasons. First, reductions in immigration into the United States, and in internal migration to cities, eased the growth in demand for housing, especially in urban areas. Second, construction activity was buoyant. Together, these reduced the tightness of the housing market and created viable alternatives to lodging. Third, a variety of new forms of social assistance reduced the effects of death, disability, or unemployment on family well-being. In a sense, this provided a substitute for lodgers' rent, reducing the incentive for families to supply lodging. Fourth, married women increasingly took jobs outside the home, reducing the attractiveness of being a landlady. This also served to reduce the supply of lodging. Finally, Modell and Hareven argue that the notion of privacy was redefined. The demand for separate accommodation grew at the expense of lodging. Families increasingly preferred, and could afford, to live alone.

This again raises the role of increasing affluence in postwar household formation. As a family's income increases, it typically has more housing alternatives. More expensive housing forms become feasible. To what extent did the demand for (or supply of) lodging decline as a consequence of rising affluence? This is a question to which we return in more detail in chapter six.

The above discussion omits, however, some additional considerations of importance. It views lodging as a commercial enterprise. However, the majority (over 70 per cent) of family and nonfamily persons in postwar Canada who did not maintain their own dwelling lived with relatives. The taking in of relatives often cannot be viewed simply as a commercial decision. Sometimes the decision to take in a relative is based strictly on a sense of familial duty, e.g. helping out someone in need of financial or other assistance. At other times, lodging represents a form of mutual assistance, with each party helping out in its own way. One example of this would be the young family that takes in a grandparent who looks after the children while the parents work. We know little about the factors shaping such lodging or about how they have changed over time.

HISTORICAL STUDIES

To understand why and how Canadians came to live in smaller households in the postwar period, it is important to know when the trend began. Was it strictly a postwar phenomenon, or did it have its origins in this century or even in the last?

It is difficult to pinpoint an exact starting date because of the paucity of data on household formation in Canada before 1881. In addition, available data for 1881 through 1941 are not exactly comparable with postwar definitions. After adjusting for definitional change, Wargon (1979a, 33) concludes that average household size has declined since at least 1931. In *Historical Statistics of Canada*, 2nd ed (Series A249), unadjusted household size data show a steady decline – from 5.3 persons in 1881, to 4.8 persons in 1911, to 4.3 persons in 1941. Kobrin (1976a) suggests a decline in the United States from 1790 to 1900, but no estimates are presented for intervening years. Did the U.S. decline begin before 1790 or between 1790 and 1900?

One view is that the modern decline in average household size began during the industrial revolution. It is beyond the scope of this book to discuss the variety of social and economic processes that are thought to have started this. The interested reader is instead referred to such works as Laslett and Wall (1972), Shorter (1977), and Levine (1977). Industrialization, it is argued, resulted in smaller family sizes through its effects on fertility, age at marriage, and living arrangement decisions. As Laslett (1969, 200) notes: "The association of the nuclear family, and therefore the small household, with industrial society and industrial society alone has been very strong. It is only now coming under criticism."

Laslett (1969) traced mean household sizes in England from the sixteenth to the nineteenth century, using detailed local census data from a sample of parishes. Some of these were rural parishes, others were from small towns or cities, and still others were within London itself. This breadth of geographic coverage is important in establishing the generality of his results. Laslett notes a problem in comparing these local census data. Virtually none of these census records contains a definition of a household. Since each local census was undertaken separately, it is possible that the use of the terms *dwelling* and *household* varied from one to the next. Having no alternative, Laslett ignored this problem. What Laslett found was controversial. In his words (200): "Mean household size remained fairly constant at 4.75 or a little under, from the earliest point for which we have found figures, until as late as 1911. There is no sign of the large, extended co-residential family group of the traditional peasant world giving way to the small, nuclear, conjugal household of modern industrial society." This suggests that modern decline in average house-

TABLE 33
Percentage distribution of households by size

Persons in household	*Canada*			*Laslett's 100 parishes*	*England and Wales*		
	1941	*1951*	*1961*	*1574–1821*	*1911*	*1951*	*1961*
1	6	7	9	6	5	11	12
2	18	21	22	14	16	28	30
3	20	20	18	16	19	25	23
4	18	19	18	16	18	19	19
5	13	13	13	15	14	10	9
6	9	8	8	12	10	4	4
7	6	5	5	8	7	2	2
8 or more	10	7	6	13	9	2	1
Total	100*	100	100	100	100	100	100

SOURCE: See Table 19 and Laslett and Wall (1972, 143).
* Columns may not total due to rounding.

hold size, in England at least, began only in the twentieth century and that it was preceded by three centuries of stable household sizes.

The pre-industrial distribution of household sizes found by Laslett resembles the distribution for Canada in 1941 despite the time difference. See Table 33. In both cases, about 6 per cent of households contained just one person. Although Canada had relatively more two- and three-person households, and relatively fewer households of six or more persons, the similarities are striking.

In later work, Laslett and Wall (1972, 1–85) compare their findings for England with other communities in France, Serbia, Japan, and the United States in the seventeenth and eighteenth centuries. They find similar mean household sizes, ranging up only to 5.85 persons in Bristol, Rhode Island, in 1689. These results suggest that English household sizes are not unlike those found in other countries as well. Again, however, one must remember that variations in the definition of a household make exact comparisons problematic.

The larger mean size reported for the American community of Bristol was augmented by the findings of Greven (1972) and Kobrin (1976a). They report averages of about six persons per dwelling in various communities in colonial America. Thus, Canada's colonial mean household sizes may also have been larger than those reported for England. Although mean household sizes have shrunk in both countries, Canada appears to have lagged well behind England and Wales in the decline of larger households. Shorter (1977, 30–1) notes the almost complete absence of extended family units in Laslett's English samples and argues that, in colonial America, such households may have been more common. If so, there may have been a decline in mean household size in the nineteenth century in the United States and Canada as these extended family households disappeared. This could explain the somewhat earlier decline in mean household size in Canada vis-à-vis England and Wales.

The studies cited above are fragmentary, and the data employed typically limit comparisons. Nonetheless, they do allow us to sketch some features of the modern decline in household size. First, it appears that this decline began substantially before 1945, possibly even before the turn of the century. Second, larger households, and in particular extended family units, seem to have been more prevalent in North America than in England or other parts of northwest Europe. The decline in mean household size in Canada may thus have preceded the trend in England – perhaps beginning even before the 1880s – as the incidence of extended family households declined. Kobrin (1976a, 127) concludes: "Early explanations attributed this fall in size to the breakup of an extended family system . . . This line of argument has generated considerable controversy . . . and the term 'extended family' still has no widely agreed-upon meaning . . . If the definition of extension requires more than one married pair . . . neither early rural nor more recent urban population in the United States give evidence of such a pattern . . . However, if 'extension' . . . mean[s] the . . . inclusion in families of adults who are not currently married – the grown children, siblings, and parents of the married couple – then extension has declined and . . . 'nuclearization' has occurred."

CONCLUSIONS

This chapter began by noting the decline in average household size over the postwar period. Prewar data on household formation indicate that this decline continued a trend that began in Canada in the 1880s or before. While this postwar trend was similar, the causes changed. As with the United States, Canada had a shortage of housing throughout much of its colonial history, largely in the form of older, less expensive dwellings. The housing stock was typically new and expensive. In part, rapid overall population growth necessitated much new construction. In part also, internal migration resulted in a continued westward movement within Canada, creating demands for housing in areas that had little or no existing stock. Over time, the housing stock matured, enabling a range from newer, more expensive housing to older, less expensive accommodation. The kinds of households that are attracted to such housing are typically nonfamily individuals with the relatively low incomes and poor (young, elderly, or lone-parent) families. In colonial times, it was these persons who could not find afforable housing. Consequently, they tended to coreside with other persons who were better off, typically middle-aged husband-wife families. Because of this, typical household sizes were larger in Canada than in northern Europe, where there was a more plentiful stock of older housing.

At least through the first half of the twentieth century, part of the decline in average household size was a kind of catch-up. With the growth of the older, less expensive housing stock, more individuals and families found separate accommodation affordable. Although the formation of these typically small households lagged behind regions such as Europe, the gap shrank. Was this, however, strictly the result of the growth of the older housing stock? I suspect not. Also important was the rising affluence of nonfamily individuals. Increasingly, they could afford separate accommodation, even newer, more expensive dwellings. In addition, the rise of assisted housing, particularly for senior citizens and other low-income groups, helped to make separate accommodation affordable.

In the postwar period, there continued to be an undoubling of families and of persons not forming part of a nuclear family. In the early 1960s, and possibly before, changes in household size largely reflected this. However, after the mid-1960s, the fall in household size began also to reflect declining fertility. Over two-thirds of all households were nuclear families living alone, and they shrank from 4.0 persons per household in 1966 to 3.3 in 1981. This decline in fertility was thus an important determinant of changing household size.

In addition, the rapid formation of households reflected the increasing number of nonfamily individuals. Changes in marriage, divorce, and separation were important, as were the effects of greatly increased longevity among women.

This chapter has illustrated the diverse and changing forces that shaped postwar household formation. The next two chapters attempt to estimate the numerical importance of several of these.

CHAPTER FIVE

Analysis of Household Formation

Commonly, household formation is analysed using headship rates. The strength of this approach is that it decomposes a change in household formation into two components: changes in population by age, sex, and marital status cohorts and changes in the propensity for members of a cohort to head a household. The latter propensity is simply the headship rate. This decomposition can be used to estimate the impact of demographic changes, such as the postwar baby boom, increasing longevity, or the marriage bust, on living arrangements assuming that headship rates do not change. Such estimates are described in this chapter as the direct effects of demographic change. They represent "what if" calculations; showing what the effect of a demographic shift might have been if the propensities to form particular living arrangements had not changed.

From chapter four, however, it is evident that headship rates did not remain constant. Headship rates increased substantially overall, and especially among the young, the elderly, and nonfamily individuals. Several explanations have been discussed already. It is difficult, however, to assess the relative importance of any of these explanations using headship rate data alone. To what extent, for example, was the rise in headship rates a result of the increasing propensity for families to live alone?

In this chapter, a components of change analysis is introduced that decomposes a headship rate into a number of components. A change in headship rate is then related to changes in components - hence, a "components of change" analysis. Components of change approaches have been used before in the analysis of household formation. See Beaujot (1977), Kobrin (1976a), and Sweet (1984), who use similar, though not identical, methods to look at household formation.

The approach taken is exemplified as follows. Suppose that there are H husband-wife families living alone (i.e. forming a husband-wife PNF household in the terminology of chapter two). Further, suppose that there

FIGURE 6

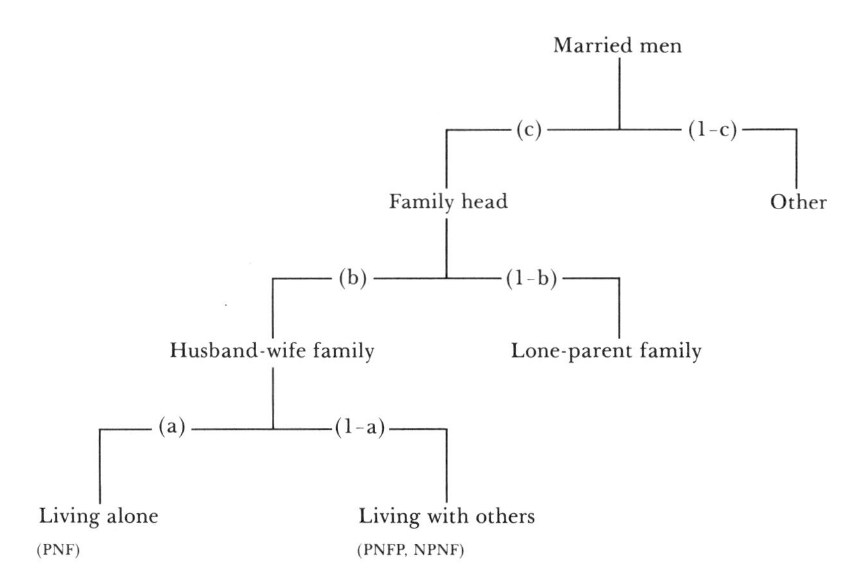

are N married males in the population, for a headship rate of $v = H/N$. This headship rate, in turn, can be expressed as the product of three components (a, b, and c) as follows:

$$v = abc, \tag{1}$$

where c is the propensity for a married male to head a family,
b is the propensity for a family head in that cohort to head a husband-wife family, and
a is the propensity for a husband-wife family headed by a person in that cohort to be living alone (i.e. as a PNF household).

Figure 6 shows that (1) is an identity. Therein, married men are divided into those heading families and other (i.e. nonfamily individuals). The proportions falling into each category, c and $1-c$ respectively, are shown in parentheses. Of the family heads, a proportion b head husband-wife families; the remainder head lone-parent families. Finally, of heads of husband-wife families, a proportion a live alone (i.e. form PNF households). The remainder live in shared accommodation.

In the postwar period, c was close to 1.0, but it declined later as the incidence of separation increased. Most male family heads headed husband-wife units, and so b was also close to 1.0. However, the relative number of fathers heading lone-parent families declined, and so b rose closer to unity. Finally, most husband-wife families lived alone; hence a

was large. As noted in chapter four, the incidence of husband-wife families living alone (i.e. *a*) rose over the postwar period. Thus, the postwar increase in headship rate can be decomposed into increases in *a* and *b* that were partly offset by a decline in *c*.

In this chapter, such a components of change approach is applied to aggregate household formation in Canada between 1971 and 1976. The use of such a brief period is made necessary by data limitations. Nonetheless, the approach is quite valuable. The first section describes data sources and limitations. Concepts underlying the components of change method are discussed in the next section. In the third section, the components are defined. Direct and interaction effects are discussed in the fourth section. The remaining sections discuss the application of this components of change approach to aggregate household formation and contrast the formation of small and large households.

DATA SOURCES

A components of change approach requires extensive cross-classified data on households, families, and individuals. Published census reports do not provide this kind of detail. Instead, the 1971 and 1976 census public use samples are used.[1] Definitional changes render problematic use of the 1981 Samples, and such samples are not available for earlier censuses. In reading this chapter remember therefore that the empirical analysis is restricted to the first half of the 1970s. Factors that may have been more important in shaping household formation at other times are not considered.

For each of 1971 and 1976, there are, in fact, three public use samples. The first is a 1 per cent sample of individuals; the second, a 1 per cent sample of families; and the third, a 1 per cent sample of households. This allows us to look at relations among the demographic characteristics of an individual, that person's family status, and household formation. The components of change method described below makes use of all three samples.[2]

For the most part, 1976 public use sample data are comparable in definition to their 1971 counterparts. We have already noted, however, that the 1971 Census excludes women from heading a family or household if her spouse is present, a restriction that was lifted in 1976. As a consequence, the 1976 Census public use samples for families and for individuals do not identify the head of a husband-wife family. They do identify, however, married (spouse present) women who are household heads. This latter group, by definition, must also head a husband-wife family. Thus, the two censuses differ insofar as one cannot identify married (spouse

FIGURE 7
Census definitions of family status of individuals

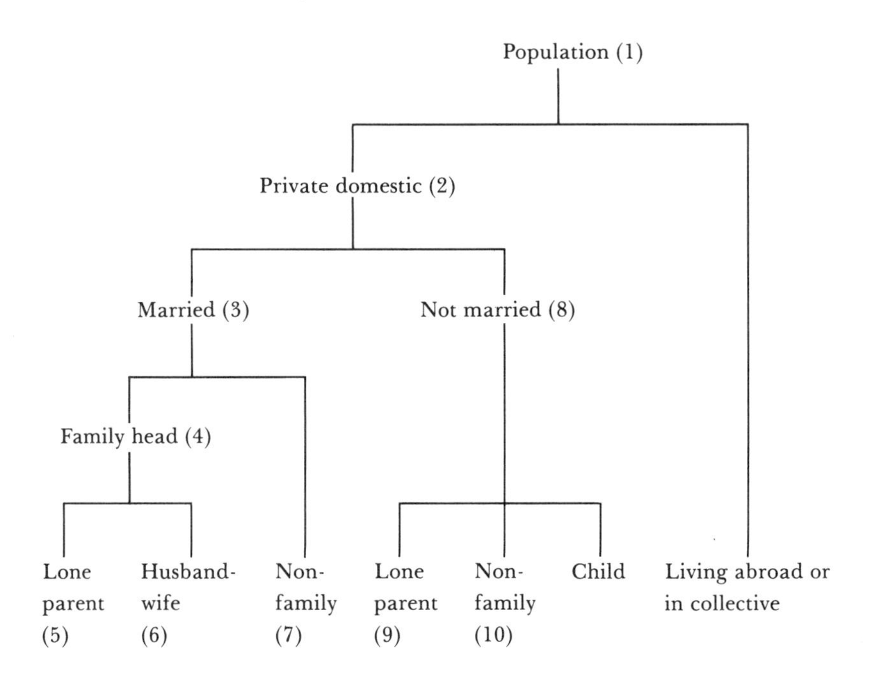

present) women who head lodging (i.e. other nonprimary) families in the 1976 samples. Other than this, the samples are quite comparable.

CONCEPTS

This section begins with a discussion of census concepts of the individual, family, and household. These different concepts are then used to define a set of propensities for the components of change analysis.

In Figure 7 is presented a schematic of the definitions for an individual's family status. At the top is the entire population in a cohort. This is divided into private individuals (those in private dwellings) and collective individuals (living in group quarters such as barracks, hospitals, or hotels). Those living in private dwellings include both domestic residents and persons living temporarily abroad in military or foreign service.[3] Domestic, private individuals can further be dichotomized on the basis of marital status: married (including separated) or not married (i.e. single, widowed, or divorced). Further, each of these can be split according to family status. Married persons can head a lone-parent or husband-wife family; otherwise, they are nonfamily persons.[4] A not-married individual

FIGURE 8
Living arrangements of families and individuals

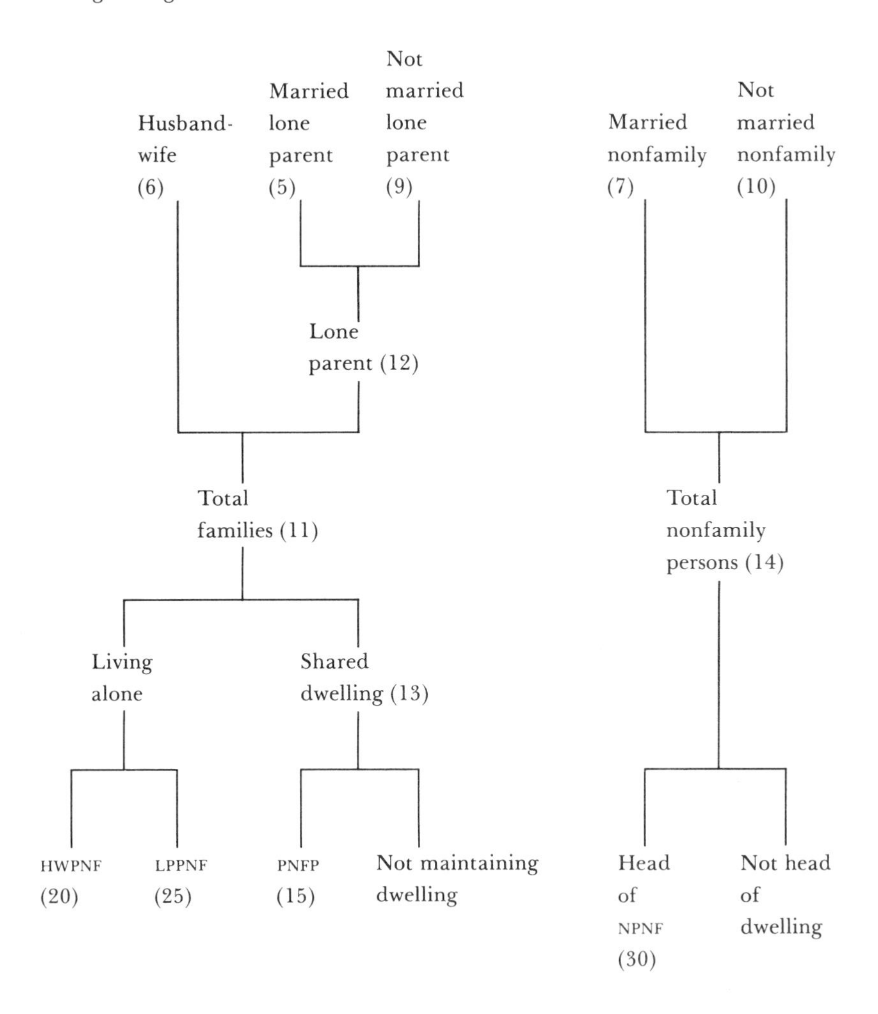

can be a lone parent, a child, or a nonfamily person. The census counts as children only never-married persons living with at least one parent. Thus a married, widowed, or divorced person can never be a "child."

These concepts underlie both the 1971 and the 1976 censuses. There are, however, some procedural differences between the two. As mentioned before, the 1976 Census allows either spouse to head a husband-wife family, whereas the 1971 Census allows only the husband. Also, the 1976 Census treats common law partners as married, whereas the 1971

Census does not. Thus, a common law couple with children might have been counted as a lone-parent family in 1971 but as a husband-wife family in 1976.

Figure 8 continues the breakdown in Figure 7 by examining the living arrangements of families and individuals. The two lone-parent categories (for married and not-married heads) are first aggregated and then added to the husband-wife families to yield total families. Then this total is disaggregated by whether the family lives alone or in shared accommodation. We distinguish among those living alone between husband-wife (HWPNF) and lone-parent (LPPNF) units. We distinguish among families in shared accommodation, between primary nuclear families (PNFP) and others. Finally, we distinguish among nonfamily individuals between those that head a household (by definition an NPNF household) and those that do not.

PROPENSITIES

In Table 34 are presented data for men aged 15–24 for 1971 and 1976. Each row is a category; where equivalent, the corresponding row numbers are shown in parentheses in Figures 7 and 8. In addition, a breakdown of households by size is shown for each of the four household types. Note that a PNFP household must contain at least three persons, and HWPNF and LPPNF households at least two. The columns marked "Number" for 1971 and 1976 give counts in hundreds of men aged 15–24, or of families or households headed by such persons. For example, there were 236,900 HWPNF households headed by a man 15–24 in 1971; by 1976, this had risen to 285,500.

Between 1971 and 1976, there were substantial increases in some of the counts and declines in others. To analyse these changes, let us define the following propensities. Each propensity p is subscripted by the row to which it refers. For example, $p[30]$ is a propensity to form NPNF households. Let $n[i]$ be the count in row i in a particular year; for the moment, we need not distinguish between 1971 and 1976:

propensity to be private, domestic individual:

$$p[2] = n[2]/n[1]; \tag{2}$$

propensities to be married or not-married:

$$p[3] = n[3]/n[2],\ p[8] = n[8]/n[2];\ \text{note}\ p[3] + p[8] = 1; \tag{3}$$

TABLE 34

Components of change for men aged 15 to 24, Canada, 1971-6

Row item	Propensity ratio to	1971 Number (00s)	1971 Propensity	1976 Number (00s)	1976 Propensity	Annual change in propensity (%)
1 Population		19,867		22,584		
2 Private, domestic	1	19,428	0.978	21,895	0.970	−0.2
3 Married	2	3,055	0.157	3,618	0.165	1.0
4 Family head	3	2,844	0.931	3,427	0.947	0.3
5 Lone parent	4	9	0.003	13	0.004	3.7
6 Husband-wife	4	2,835	0.997	3,414	0.996	−0.0
7 Nonfamily	3	211	0.069	191	0.053	−5.2
8 Not married	2	16,373	0.843	18,277	0.835	−0.2
9 Lone parent	8	26	0.002	11	0.001	−17.6
10 Nonfamily	8	2,234	0.136	2,782	0.152	2.2
11 All family heads		2,870		3,438		
12 Lone parents		35		24		
13 Families sharing acc	11	453	0.158	397	0.116	−6.1
14 Nonfamily persons		2,445		2,973		
15 PNFP	13	218	0.481	202	0.509	1.1
16 3 persons	15	71	0.326	80	0.396	4.0
17 4 persons	15	67	0.307	70	0.346	2.4
18 5 persons	15	45	0.206	33	0.163	−4.6
19 6+persons	15	35	0.161	19	0.094	−10.1
20 HWPNF	6	2,369	0.836	2,855	0.836	0.0
21 2 persons	20	1,224	0.517	1,715	0.601	3.1
22 3 persons	20	832	0.351	823	0.288	−3.9
23 4 persons	20	263	0.111	273	0.096	−2.9
24 5+persons	20	50	0.021	44	0.015	−6.1
25 LPPNF	12	14	0.400	5	0.208	−12.2
26 2 persons	25	4	0.286	5	1.000	28.5
27 3 persons	25	6	0.429	0	0	
28 4 persons	25	2	0.143	0	0	
29 5+persons	25	2	0.143	0	0	
30 NPNF	14	671	0.274	1,166	0.392	7.4
31 1 person	30	351	0.523	706	0.606	3.0
32 2 persons	30	206	0.307	319	0.274	−2.3
33 3 persons	30	74	0.110	95	0.082	−5.9
34 4 persons	30	20	0.030	32	0.027	−1.6
35 5+persons	30	20	0.030	14	0.012	−16.6

SOURCE: Computed from the 1971 and 1976 census public use sample tapes. See text for details.

*Estimates based on 1 per cent samples excluding Prince Edward Island, Yukon, and Northwest Territories propensities for married person to head family or be nonfamily person:

$$p[4] = n[4]/n[3], p[7] = n[7]/n[3]; \tag{4}$$

propensities for married family head to head husband-wife or lone parent unit:

$$p[5] = n[5]/n[4], p[6] = n[6]/n[4]; \text{ note } p[5] + p[6] = 1; \tag{5}$$

propensities for not-married person to be lone parent or nonfamily person:

$$p[9] = n[9]/n[8], p[10] = n[10]/n[8]; \tag{6}$$

propensity for family to share accommodation:

$$p[13] = n[13]/n[11]; \tag{7}$$

propensity for family sharing accommodation to be primary family:

$$p[15] = n[15]/n[13]; \tag{8}$$

propensities for PNFP households to be of certain size:

$$p[16] = n[16]/n[15], p[17] = n[17]/n[15], p[18] = n[18]/n[15], p[19] = n[19]/n[15]; \text{ note } p[16] + p[17] + p[18] + p[19] = 1; \tag{9}$$

propensity for husband-wife family to live alone:

$$p[20] = n[20]/n[6]; \tag{10}$$

propensity for HWPNF household to be of certain size:

$$p[21] = n[21]/n[20], p[22] = n[22]/n[20], p[23] = n[23]/n[20], p[24] = n[24]/n[20]; \text{ note } p[21] + p[22] + p[23] + p[24] = 1; \tag{11}$$

propensity for lone-parent family to live alone:

$$p[25] = n[25]/n[12]; \tag{12}$$

propensity for LPPNF household to be of certain size:

$$p[26] = n[26]/n[25], p[27] = n[27]/n[25], p[28] = n[28]/n[25], p[29] = n[29]/n[25]; \text{ note } p[26] + p[27] + p[28] + p[29] = 1; \tag{13}$$

propensity for nonfamily person to head household:

$$p[30] = n[39]/n[14]; \quad (14)$$

propensity for NPNF household to be of certain size:

$$p[31] = n[31]/n[30],\ p[32] = n[32]/n[30],\ p[33] = n[33]/n[30],\ p[34] = n[34]/n[30],\ p[35] = n[35]/n[30];\ \text{note}\ p[31] + p[32] + p[33] + p[34] + p[35] = 1. \quad (15)$$

Values for these propensities in 1971 and 1976 are shown in the two columns in Table 34 titled "Propensity." The rate of change in each propensity between 1971 and 1976 is shown in the right-hand column. Note, for example, the increase in the propensity for young HWPNF households to be of size 2 [i.e. childless] at the expense of larger households, reflecting the decline in, or delay of, fertility between 1971 and 1976.

Given the propensity values described above, and the total population of an age-sex cohort, all of the counts described in Table 34 can be reproduced recursively. For row i, the count $n[i]$ is simply $p[i] * n[j]$, where $n[j]$ is the count for the row that forms the denominator of $p[i]$. For example, $n[2] = p[2] * n[1]$ and $n[35] = p[35] * n[30]$. The only exceptions are rows [11], [12], and [14], for which no propensities have been defined. For these:

$$n[11] = n[4] + n[9],\ n[12] = n[5] + n[9],\ n[14] = n[7] + n[10]. \quad (16)$$

DIRECT AND INTERACTION EFFECTS

Consider the number of one-person households formed by men aged 15-24. It is seen from Table 34 that there were 35,100 such households in 1971, and 70,600 five years later. Using the propensities defined above, such counts can be expressed as follows:

$$n[31] = p[31] * p[30] * (p[7] * p[3] + p[10] * (1\text{-}p[3])) * p[2] * n[1], \quad (17)$$

where $n[1]$ is the total population cohort, $p[2]$ is the propensity to reside in a domestic, private household, $p[3]$ is the propensity to be married, $p[7]$ and $p[10]$ are the propensities for married and not-married persons respectively to be nonfamily individuals, $p[30]$ is the propensity for a nonfamily person to head a household, and $p[31]$ is the propensity for that household to be of size 1.

This decomposition allows us to measure the impact of the change in any one component. For example, what effect did the changing propensity to be married have on the formation of one-person households between 1971 and 1976? Imagine that, between 1971 and 1976, none of the other components had changed except for the marriage propensity. From equation (17), we can predict the number of households that would have been present in 1976. The difference between this projection and the actual number of 1971 one-person households is the direct effect of the changing marriage propensity.

Similarly, one can measure the direct effect of the change in any other component. Allowing for the fact that simultaneous changes must be considered where propensities sum to 1 (e.g. $p[3]$ and $p[8]$), direct effects can be estimated for 14 different propensities or sets of propensities. See Table 35. In addition, the direct effect of population growth can be estimated.

These direct effects do not sum, however, to the total change in household formation between 1971 and 1976. This can be illustrated using the following argument. Suppose, for simplicity, that

$$H = abN, \tag{18}$$

where H is the number of households, N is the number of persons in a cohort, and "a" and "b" are propensities. If, between 1971 and 1976, "a" increases by a proportion "r," "b" by "s," and N by "v," the increase in households ($H_{76} - H_{71}$) is:

$$H_{76} - H_{71} = ((r + s + v) + (rs + rv + sv + rsv))a_{71}b_{71}N_{71}. \tag{19}$$

The term $(r + s + v)$ on the right hand side gives the sum of the direct effects of the propensities a and b, and population growth. The remaining term, $(rs + rv + sv + rsv)$ is the interaction effect: i.e. the impact of the simultaneous changes in a, b, and n.

In Table 35 are presented the calculated direct and interaction effects for males aged 15–24 in Canada. The right-hand column gives the net number of new households contributed by the change in a component (one of the 14 propensities or population growth), and the remaining columns disaggregate these by household size. The bottom row shows that between 1971 and 1976 the number of households headed by young men increased by 95,600. Virtually all of this increase was in one- and two-person households.

In part, the increase was attributable to population growth. During this period, the population of this age cohort swelled by 14 per cent with the entry of the baby boomers. Population growth alone, in the absence of change in any propensity, would have generated 44,700 more house-

TABLE 35
Direct and interaction effects (in hundreds) for men 15 to 24, Canada, 1971-6

Row	Item	*Households by number of persons**					*All hhlds*
		1	*2*	*3*	*4*	*5+*	
	Direct effects						
1	2	-3	-12	-8	-3	-1	-27
2	3,8	-2	61	46	17	6	128
3	4,7	-7	17	14	5	2	31
4	5,6	0	-1	0	0	0	-1
5	9,10	37	20	5	1	1	64
6	13	0	0	-19	-18	-21	-58
7	15	0	0	4	4	5	13
8	16,17,18,19	0	0	15	9	-24	0
9	20	0	1	1	0	0	2
10	21,22,23,24	0	199	-149	-36	-13	0
11	25	0	-2	-3	-1	-1	-7
12	26,27,28,29	0	10	-6	-2	-2	0
13	30	151	88	32	9	9	289
14	31,32,33,34,35	55	-22	-19	-2	-12	0
23	Sum of above	231	360	-89	-18	-53	431
24	Population growth	48	196	134	48	21	447
25	Total direct effects	279	556	46	30	-32	879
26	Interaction effects	76	49	-31	-7	-10	77
27	Total change 1971-6	355	605	15	23	-42	956

SOURCE: Computed from data in Table 34. See text for details.
* See note to Table 34.

holds. See row 24 in Table 35. In other words, the increased size of the population cohort itself accounts for about one-half of the net formation of households between 1971 and 1976.

What accounts for the other half? The direct effects of the change in each propensity are shown in the first 14 rows of the right-hand column. Of these, the direct effect of the increase in $p[30]$ is the largest. In other words, the increased propensity for young nonfamily men to head their own household would have accounted for 28,900 additional households in the absence of change in any other component. The next largest effect is attributable to the shift in $p[3]$ (and hence $p[8]$). The increased propensity for young men to be married itself accounted for 12,800 additional households between 1971 and 1976. Altogether, the direct effects of these and other propensity shifts accounted for almost one-half of the net formation of households in this age group.

To this point, the discussion has not considered the growth of particular sizes of household. In Table 35, the left-hand columns give the direct and interaction effects for each of five size categories.

Consider, first, the one-person household. Between 1971 and 1976, the number of such households headed by men 15–24 increased by 35,500. The direct effect of population growth was, however, only 4,800, less than 14 per cent of the observed increase. In other words, population growth itself was not an important determinant of the formation of one-person households. The largest direct effect (15,100 households) was the increase in $p[30]$; it alone accounted for about one-half of the total increase in one-person households.

Population increase was also unimportant in the growth of two-person households, having a direct effect of 19,600 units out of a 1971–6 net change of 60,500. As important as population growth was the shrinking typical size of an HWPNF household. In particular, the rising propensity for young HWPNF households to be childless (hence of size 2) accounted for 19,900 more two-person units between 1971 and 1976. Also important was the increasing propensity for young nonfamily men to form NPNF households (see row 13).

Among households of size 3 or larger, the direct effects of population growth are strongly to weakly positive. However, the actual increase in numbers of such households ranged from weakly positive to weakly negative. In general, the positive direct effect of population growth was offset by the negative direct effects of other components. The most important of these was the rising incidence of childless HWPNF (see row 10), i.e. a decline in the propensity for HWPNF to form households of three or more persons. Among five-plus-person households, other direct effects were equally important, but the magnitudes involved are small.

These direct effects are powerful tools in assessing the empirical importance of different reasons for the changing rate of household formation. For example, in chapter one, it is suggested that part of the postwar household formation boom was attributable to earlier home leaving among young adults. When a person leaves the parental home, he or she becomes a nonfamily person. Thus, earlier home leaving should manifest itself as an increase in $p[10]$. Between 1971 and 1976, $p[10]$ did increase for young not-married men, from 0.136 to 0.152 in Table 34. However, the direct effect of this was not large – only 6,400 net additional households, or less than 7 per cent of net increase in households headed by young men. Wargon (1979a, Table 3) finds similar evidence that changes in home leaving patterns among young Canadians were relatively minor over much of the postwar period.

Chapter one notes also the decline of shared accommodation as a reason for greater household formation. This decline in shared accommodation is reflected in several propensities. One is $p[30]$. Young nonfamily men may lodge, share a dwelling with partners, or live on their own. A rise in $p[30]$ is consistent with a decline in lodging and an increasing preference to live alone or with a smaller group of partners. The decline in shared accommodation is also reflected in the smaller values for *p*[13], *p*[18], and *p*[19]. However, the direct effects of these latter three are considerably smaller than for *p*[30]. In other words, the undoubling of family households was not nearly as important a source of household formation in the early 1970s as was the undoubling of nonfamily individuals into separate accommodation.

It is also argued in chapter one that changing marital preferences account for some of the changing pattern of household formation. In chapter three, it is argued that the 1970s saw the beginning of the marriage bust, characterized by a delay or forgoing of marriage. The data in Table 34 do not show a marriage bust among men aged 15–24 between 1971 and 1976. The propensity to be married actually rose slightly, although other evidence points to a decline after 1976. The direct effects of this rise in *p*[3] were substantial and positive, especially for two- and three-person households.

We have also considered the argument that changing fertility patterns, in particular the post-1960 baby bust, reshaped the size distribution of households. This shows up clearly in row 10 of Table 35. The increasing propensity for young HWPNF households to be childless contributed substantially to the growth of two-person units and to the decline of larger units.

AN ANALYSIS OF TOTAL HOUSEHOLD FORMATION

Analyses of the type summarized in Tables 34 and 35 were undertaken for each of ten age groups by two sexes. Results are summarized in Table 36. The right-hand column shows net aggregate household formation between 1971 and 1976 by age and sex of household head, as computed from the 1971 and 1976 public use samples. The bottom right-hand value, for example, indicates that the number of households in Canada increased by about 1.1 million over the five-year period.

The left-hand columns of Table 36 show the decomposition of this net household formation into direct and interaction effects. The bottom row presents column totals. The direct effect of population growth, for example, was 688,000 additional households, just over 60 per cent of total net household formation. In other words, just under 40 per cent of net household formation was accounted for by the direct effects of changing propensities or interaction effects. Using a similar approach, Sweet (1984) found that two-thirds of U.S. net household formation between 1970 and 1980 was attributable to a changing mix of population by age, sex, and marital status. In Table 36, marital status changes are handled separately

TABLE 36

Components of change in total household formation (in 000s): Canada, 1971-6

	*Direct effect**												
	Propensity											*Inter-action effect*	*Net hhld form'n 1971–6*
	2	*3,8*	*4,7*	*5,6*	*9,10*	*13*	*15*	*20*	*25*	*30*	*Pop'n growth*		
(a) Male heads													
Total	−19	19	20	3	11	−97	−4	85	7	120	544	11	700
15-24	−3	13	3	0	6	−6	1	0	−1	29	45	8	96
25-29	−2	−8	0	1	3	−11	1	0	−1	19	144	3	150
30-34	−3	−1	1	1	2	−13	−1	6	0	12	127	2	133
35-39	0	0	4	0	2	−15	−5	42	2	8	12	2	52
40-44	−1	6	3	0	0	−17	6	−29	1	10	19	−6	−8
45-49	−2	2	2	1	−2	−7	−9	7	4	6	16	0	18
50-54	−1	−1	2	0	−1	−5	−1	11	−1	8	54	0	66
55-59	−2	1	1	0	0	−10	6	15	1	4	12	−1	26
60-64	−1	1	1	0	0	−3	−4	6	4	4	45	0	53
65+	−5	6	2	0	0	−10	1	27	−1	21	71	3	114
(b) Female heads													
Total	−6	−6	50	−28	36	−22	23	0	16	156	143	66	429
15-24	0	0	8	−4	11	−2	3	0	2	32	10	17	77
25-29	0	4	8	−5	4	−1	2	0	6	23	14	26	81
30-34	0	4	7	−4	3	0	1	0	−1	8	12	10	38
35-39	0	6	8	−2	6	−3	3	0	−2	3	3	2	25
40-44	0	2	3	−3	3	−1	1	0	2	8	−1	3	18
45-49	0	−1	4	−3	4	−6	12	0	1	4	2	−5	12
50-54	0	−2	4	−3	2	−3	−1	0	5	5	14	3	23
55-59	0	−5	3	−1	1	−2	1	0	0	14	7	2	21
60-64	0	−5	1	−1	1	0	−1	0	2	8	24	3	30
65+	−5	−9	5	−2	1	−4	2	0	0	51	59	4	103
Total	−25	14	70	−25	47	−119	19	85	23	277	688	76	1,129

SOURCE: Computed from the 1971 and 1976 census public use samples. See text for details.

* See note to Table 34.

(in *p*[3] and *p*[8]); hence the direct effect of population change estimated for Canada should be somewhat smaller. In this respect, Sweet's finding is broadly similar to that for Canada.

Which propensity had the largest direct effect? The increasing propensity for nonfamily persons to head a household, *p*[30], alone accounts for about one-quarter of net household formation. Among the propensities, it had by far the largest direct effect. Other important propensity changes were the decline in families sharing accommodation (*p*[13]) and the rise in husband-wife families living alone (*p*[20]). The direct effect of *p*[13] is negative, indicating a decline in the formation of PNFP households, while the direct effect of *p*[20] is positive. In fact, the net impact of the decline in shared accommodation shows up principally as a rise in *p*[30], i.e. nonfamily persons choosing to form their own household rather than lodge with families or partners.

Changes in marital status (*p*[3] and *p*[8]) had little effect on net household formation. The 1970s did mark the beginning of the marriage bust. For women under 45, and men between 25 and 34, *p*[3] did decline from 1971 to 1976. However, the decline was modest, and the direct effect on net household formation was correspondingly small.

The direct effect of a rising *p*[30] was larger among the young and the elderly. Because most middle-aged persons were married, relatively few were nonfamily individuals. This limited formation of NPNF households in these age groups. However, many young adults were single, and there was a substantial number of elderly widows. It was among these groups that the rising incidence of headship was most pronounced. In other words, for some reason, young singles and elderly widows became much more likely to head a household.

We can use Table 36 to estimate the relative empirical importance of different explanations of postwar household formation. In chapter one, we considered ten explanations. How important were these in shaping household formation between 1971 and 1976? Let us consider each in turn.

The postwar baby boom. The leading edge of the baby boom was about 30 years old in 1976. Between 1971 and 1976, Table 36 indicates that the direct effect of population growth among the under-30s was over 200,000 households, or about 20 per cent of total net household formation during this period.

Changing life cycles. There have been changes in typical age at marriage and in the number and spacing of children borne. However, it is difficult to tell what effects these changes have had on aggregate household formation from this components of change analysis. See marriage rush and marriage bust below.

Increased longevity. As discussed in chapter three, there were substantial improvements in longevity for men and women in the postwar period. To measure the impact of improved longevity on household formation, one should first estimate what the 1976 population might have been in the absence of reduced mortality. Such an approach is beyond the scope of this chapter. Instead, let us simply note from Table 36 that the direct effect of population growth on households headed by someone 60 or older was almost 200,000 units. Since population growth reflects historical patterns of births and immigration as well as changing mortality, it is apparent that the direct effect of increased longevity alone is likely somewhat smaller than this. Nonetheless, the impact could be of the same order of magnitude as the postwar baby boom explanation above.

Marriage rush and marriage bust. Although the early 1970s marked the beginning of the marriage bust, the direct effects of changing marital preferences on household formation tended to be negligible.

Earlier home leaving. Earlier home leaving is reflected in a rising $p[10]$. Among persons under 25, the direct effect of $p[10]$ is only 16,000 additional households, only a small proportion of aggregate household formation between 1971 and 1976.

The rising incidence of divorce. The effect of rising divorce rates on household formation was also likely small. Divorced persons are grouped with widowed and single persons as "not marrieds" in this components of change analysis, trends being evidenced as changes in $p[10]$. Although a crude way of measuring the effect of rising divorce, the direct effect of $p[10]$ on household formation was small for every age group.

The demand for privacy. As argued above, the decline in shared accommodation is reflected in both a shrinking $p[13]$ and a rising $p[30]$. Taken together, these changes account for up to 150,000 net additional households by 1976.

Increasing housing affordability. The empirical significance of this explanation cannot be assessed with the components of change analysis described in this chapter. It may well be, for example, that the rising incidence of headship, $p[30]$ reflected an increased ability to afford. We return to this affluence explanation in chapter six.

The role of the public sector. This is another explanation that cannot be assessed with a components of change analysis. To what extent, for example, was the rapid rise in household formation among the elderly due to senior citizen housing wherein rents are geared to income? We consider this issue again in chapter nine.

Improvements in home-making technology. This is a final explanation for which the components of change analysis is of little value. The effects of improved home-making technology are simply too pervasive to be isolated.

AN ANALYSIS OF ONE-PERSON HOUSEHOLD FORMATION

In chapter four, it is argued that the rise of the person living alone (i.e. the one-person household) accounted for a substantial portion of net postwar household formation. Which components of change contributed to this rise? The data for one-person households presented in Table 37 help in answering this question.[5] Between 1971 and 1976, there was a net formation of about 400,000 persons living alone. Of these, five of every ten on average consisted of persons under 35 years old and another two were elderly women (typically widowed). To what extent was such formation attributable to population growth, specifically the maturing of the postwar baby boom and increasing longevity? The components of change analysis indicates that, while population growth was not unimportant, other factors were just as, or more, important. The direct effect of population growth was less than 40 per cent of the net formation of one-person households. Contrast this with Table 36, where population growth directly accounts for 60 per cent of all household formation.

The dominant component is $p[30]$, the propensity for nonfamily persons to head households. It accounts directly for almost one-half of the formation of one-person households. In addition, the rise in $p[31]$ was important. This is the propensity for NPNF households to contain just one person. In other words, between 1971 and 1976, it became more common for nonfamily persons to head their own household and for such households to contain just one person.

It is instructive to consider briefly the direct effects of the remaining propensities. The direct effects of the propensity to be a domestic, private individual, $p[2]$, are in general negligible. However, among persons 65 or older, there was a decline in $p[2]$. This reflects in part the increased availability of group homes, nursing homes, and other institutional residence forms for the elderly. In such accommodation, the resident may have his or her own bedroom but share other rooms with the rest of the residents. Thus these residences are collectives, and residents are not included in census counts of the private population. In the absence of any other changes, the decline in $p[2]$ would have resulted in 5,000 fewer persons living alone between 1971 and 1976.

Changes in marital status did not have a substantial effect on the formation of one-person households either. In the marriage bust, $p[3]$ began to decline slightly for persons under 40 years of age. This resulted in small positive direct effects on the formation of one-person households. However, $p[3]$ was still rising among older persons, reflecting the cumulative effect of the recently ended marriage boom. The higher propensity to be married among older persons resulted in negative direct effects on the

TABLE 37

Components of change in one-person household formation (in 000s), Canada, 1971-6

	*Direct effect**								
	Propensity							*Inter-action effect*	*Net hhld form'n 1971-6*
	2	*3,8*	*4,7*	*9,10*	*30*	*31*	*Pop'n growth*		
(a) Male heads									
Total	−1	−9	−18	9	79	42	38	23	157
15-24	0	0	−1	4	15	6	5	8	36
25-29	0	1	0	2	12	3	8	8	34
30-34	0	0	−1	2	8	4	5	6	24
35-39	0	0	−3	2	6	4	0	1	10
40-44	0	−2	−3	0	6	5	1	−1	5
45-49	0	−1	−2	−1	4	5	1	0	5
50-54	0	1	−2	−1	6	4	3	1	11
55-59	0	−1	−1	0	3	4	1	0	5
60-64	0	−1	−2	1	3	1	4	0	5
65+	−1	−6	−3	0	16	6	10	0	22
(b) Female heads									
Total	−4	−7	−10	−4	111	45	75	32	240
15-24	0	0	0	3	18	6	4	10	41
25-29	0	2	0	0	15	4	5	11	38
30-34	0	2	−1	0	6	2	3	3	15
35-39	0	2	−2	0	2	1	1	1	5
40-44	0	0	−1	−2	6	2	0	−1	4
45-49	0	0	−2	−1	3	3	0	0	3
50-54	0	−1	−1	−2	3	4	5	0	10
55-59	0	−3	−1	0	11	3	4	0	14
60-64	0	−3	−1	0	6	9	14	2	26
65+	−4	−6	−1	−2	41	11	39	6	84
Total	−5	−16	−28	5	190	87	113	55	397

SOURCE: As for Table 36

* See note to Table 34.

formation of one-person households. These negative effects outweighed the positive direct effects evidenced in the younger age groups.

Between 1971 and 1976, the incidence of nonfamily status among married persons declined in every age group. When a married couple separates, prior to divorce, one spouse normally retains custody of any children while the other spouse lives elsewhere and is hence a nonfamily person. When there are no children present, both spouses become nonfamily persons upon separation. The decline in fertility between 1971 and 1976 should have meant that more (separated) marrieds were nonfamily persons. However, this tendency was offset by a higher incidence of divorce. Rather than continuing as separated but married, individuals became more likely to divorce. Consequently, the proportion of marrieds who were nonfamily persons declined. As a result, the direct effect of a changing *p*[7] was a small decline in the number of persons living alone.

Among not-married individuals (i.e. single, widowed, or divorced), the propensity to be a nonfamily person, *p*[10], also changed only slightly between 1971 and 1976. Among 15–24-year-olds, *p*[10] rose, reflecting somewhat earlier home leaving among singles. However, the direct effect was only about 7,000 net additional persons living alone. Presumably, this is because young adults found it expensive to live alone, choosing instead some form of shared accommodation. In older age groups, both positive and negative direct effects are evidenced, although these are all small.

AN ANALYSIS OF THE FORMATION OF LARGE HOUSEHOLDS

Associated with the rise of the one-person household was the decline of the large household. Households shrank in size for three principal reasons: the decline in rooming, the baby bust, and earlier home leaving among young adults. How important were these as explanations of change between 1971 and 1976?

Let us define a large household to be any unit consisting of five or more persons. Between 1971 and 1976, the number of such households dropped by 207,000 units. At the same time Canada's population increased. To be fair, much of the population increase was among young adults and the elderly, where households are typically smaller in any case. However, if there had been no change in any of the propensities in our components of change analysis, population growth alone should have generated an additional 121,000 large households, even after taking into account the age mix of the population increase. Evidently then, several of the propensities must have had large negative direct effects. Table 38 indicates the most important changes.

Note first that few households of five or more persons are headed by women. Further, there are few NPNF (head of household is nonfamily individual) or LPPNF (lone-parent family living alone) households of this size. Almost all large households are headed by men and consist of either a husband-wife family living alone (HWPNF) or a family with lodgers present (PNFP). The ensuing discussion focuses on these two types.

Numerically, the largest direct effects are attributable to *p*[24], the propensity for a husband-wife family living alone (HWPNF) to be of five or more persons. In 1976, age 40 is useful for distinguishing between two kinds of families. Families where the head was over 40 tended to have gone through their child bearing period prior to the mid-1960s. In other

TABLE 38
Components of change in formation of households (in 000s) of five persons or more, Canada, 1971-6

	*Direct effect**									
	Propensity								*Inter-action effect*	*Net hhld form'n 1971-6*
	3,8	*4,7*	*13*	*15*	*18,19*	*20*	*24*	*Pop'n growth*		
(a) Male heads										
Total	7	13	-61	-5	-32	20	-230	121	-39	-211
15-24	1	0	-2	0	-2	0	-1	2	-1	-4
25-29	-1	0	-6	1	-5	0	-16	19	-5	-14
30-34	-1	1	-10	-1	-6	2	-65	47	-17	-51
35-39	0	4	-12	-4	-4	23	-63	6	-6	-56
40-44	5	4	-14	5	0	-16	-34	10	-2	-42
45-49	2	2	-5	-6	-3	3	-14	7	-1	-15
50-54	-1	2	-3	-1	-4	4	-18	18	-3	-6
55-59	1	0	-5	3	-1	2	-6	2	-1	-5
60-64	0	0	-1	-2	-5	1	-6	6	-1	-8
65+	1	0	-3	0	-2	1	-7	4	-2	-10
(b) Female heads										
Total	1	14	-8	10	-2	0	0	8	-7	4
15-24	0	1	0	1	0	0	0	0	-1	1
25-29	0	1	0	1	0	0	0	1	0	2
30-34	0	2	0	0	-1	0	0	2	0	0
35-39	1	4	-1	1	0	0	0	1	-1	2
40-44	0	2	-1	1	0	0	0	0	1	2
45-49	0	2	-3	6	0	0	0	0	-4	1
50-54	0	1	-1	0	0	0	0	2	-1	-1
55-59	0	1	-1	0	-1	0	0	0	0	-1
60-64	0	0	0	0	0	0	0	1	0	0
65+	0	0	-1	0	0	0	0	1	-1	-2
Total	9	27	-69	5	-34	20	-230	121	-46	-207

SOURCE: As for Table 36
* See note to Table 34.

words, they parented the trailing edge of the baby boom. Families wherein the head was under 40 went through their child bearing years in the mid-1960s or later; they parented the beginning of the baby bust. The impact of declining fertility among the under-40s on HWPNF size is indirectly apparent in Table 38. It accounts for a reduction of about 145,000 large households.

However, $p[24]$ also had substantial negative direct effects among the over-40s. This may seem somewhat surprising. After all, the postwar baby boom lasted almost 20 years, during which birth rates were at a sustained high rate. Therefore, family heads aged 45, for example, in

1971 would have typically had a fertility history not unlike 45-year-olds in 1976. The decline in *p*[24], and hence its negative direct effect, cannot therefore be attributed to fertility rate changes. Instead, the explanation lies in the presence or absence of young adults in the household: in other words, home leaving. Older HWPNFs in 1976 were less likely to have as many children still living at home as their 1971 counterparts. The effect of this home leaving is substantial. The direct effect of *p*[24] was a drop of 66,000 large households in five years, among heads aged 40 to 54.

Another important factor was the decline in *p*[13], the propensity for families to share accommodation. The direct effect was 69,000 fewer large PNFP households by 1976. Much of this reflects a decline in the incidence of rooming, in the sense that fewer families were taking in roomers. To a lesser extent, it also reflects a decline in the incidence of families (typically the young and the elderly) that themselves lodge.

In such ways, a components of change analysis illustrates the relative numerical importance of different explanations for the decline of the large household. For the period 1971–6, the most important explanation was the baby bust, which accounted for a net reduction of about 145,000 large households. In addition, net reductions of over 60,000 large households each can be attributed to earlier home leaving and to the decline of lodging. The direct effects of other propensities are all considerably smaller in magnitude.

CONCLUSIONS

The components of change method identifies two kinds of components: population and a set of propensities. Changes in household formation are decomposed into the direct and interaction effects of changes in these components. This chapter has described the application of a components of change analysis applied to each of 10 age groups by two sexes between 1971 and 1976.

Overall, it is found that population growth accounted directly for about 60 per cent of the 1.1-million increase in households. Two categories of population growth were especially important. One was the result of the aging postwar baby boom. The leading edge of the baby boom was about 30 years old in 1976. Between 1971 and 1976, the direct effect of population growth among the under-30s was over 200,000 households, or about 20 per cent of total net household formation during this period. The second category was the growth of the elderly population. The direct effect of population growth on households headed by someone 60 or older also accounts for almost 20 per cent of aggregate net household formation. This provides an upper limit for estimates of the effects of increased longevity on household formation.

It is also evident, though, that population growth is not the only factor behind household formation. Just under 40 per cent of net household formation was accounted for by the direct effects of changing propensities, or by interaction effects. Which propensity had the largest direct effect on aggregate household formation?

The increasing propensity for nonfamily persons to head a household in itself accounts for about one-quarter of net household formation, i.e. more than the impact of either the maturing postwar baby boom or increasing longevity. Among the propensities, it had by far the largest direct effect.

Further, most of this involved formation of one-person households, thereby contributing substantially to the decline in average household size. The direct effect of a rising propensity for nonfamily persons to live alone was larger among the young and the elderly. Because most middle-aged persons were married, relatively few were nonfamily individuals, limiting the formation of NPNF households in these age groups. However, many young adults were single, and there was a substantial number of elderly widows. It was among these groups that the rising incidence of headship was most pronounced.

Other important propensity changes were the decline in families sharing accommodation and the rise in husband-wife families living alone. Up to about 15 per cent of net aggregate household formation between 1971 and 1976 can be attributed to the increasing propensity for families to live alone, rather than in shared accommodation.

Further, the decline of shared accommodation was an important contributor to the decline in average household size. The direct effect of the decline in shared accommodation was a reduction of 69,000 large (five-plus persons) households between 1971 and 1976. Also important in the shrinking average household size was the decline in fertility and earlier home leaving, the direct effects of which were reductions of about 145,000 and 66,000 large households respectively.

These findings reinforce the argument that postwar household formation represented the combined effects of a number of different trends. To this point, we have emphasized demographic trends. In the next chapter, we consider the role of growing affluence.

CHAPTER SIX

Income and Household Formation

Many explanations have been offered for the rapid rise in headship rates outlined in chapter four. In chapter five, the relative empirical contributions of some of these are estimated using a components of change analysis. However, one explanation that is not easily evaluated using that approach is increasing affluence. The postwar period was, in general, quite prosperous. Although not all sectors of society benefited equally, there were substantial gains in real income for most consumers. See chapter eight for more empirical detail. What role did this growing affluence play in the rise of headship rates? As incomes rose relative to the cost of housing and other goods, separate accommodation becomes more affordable. Just how sensitive were living arrangement choices to the growth of real incomes? The postwar sensitivity (or elasticity) of the demand for separate accommodation to income is the topic of this chapter.

This chapter serves several purposes. One is to identify the factors that affect the elasticity of demand for separate accommodation. We shall consider how elasticity varies with income, demographic characteristics, and geographic location. This discussion helps us to understand what kinds of consumers were most likely to form new households as incomes rose. A second purpose of this chapter is to measure the relative importance of income as a determinant of living arrangement. It has already been argued that changes in age, sex, and marital and family status composition caused changes in living arrangements. Income is another determinant. Just how important was it, though? Were the changes in living arrangement associated with rising income as substantial, for example, as those induced by changing nuptiality?

This chapter begins with a review of three major empirical contributions to this debate. All three are aggregate in approach; i.e. they look at either total household headship or just a few categories of headship. All three are also based on time series data. Although valuable insights can be gained from them, it is difficult to separate clearly the role of growing prosperity from other explanations of household formation. To do this, we turn to cross-sectional data.

The chapter then introduces two measures of the choice of living arrangement. Factors shaping the living arrangements of husband-wife families, lone-parent families, and nonfamily individuals are discussed. Relationships between these choices and income, age, family type, and family size are identified. Next are considered the data sources from which the subsequent cross-sectional evidence is derived. Three following sections consider the findings for husband-wife families, lone-parent families, and nonfamily individuals. These show how income affects living arrangement cross-sectionally. Finally, we speculate on how changes in income over time affected postwar household formation.

AGGREGATE INTERTEMPORAL ANALYSES

This section considers three aggregate studies of the economic determinants of household formation. The first, by Hickman (1974), looks at the U.S. experience between 1922 and 1970. The second, by Markandya (1983), concerns British household formation in the 1960s and 1970s. While providing valuable insights into their own nation's experiences, these studies also make useful suggestions about the analysis of Canadian household formation. The third paper, by L.B. Smith (1984), focuses on Canada in the 1960s and 1970s.

The American Experience

Hickman (1974) constructed a model of U.S. household formation between 1922 and 1970. The model is described in more detail in chapter eight. In brief, he looked at the difference over time between the actual number of households (HH) and a "standardized" number of households (HHS). He used headship rates for 1940 to estimate HHS for each year from 1922 to 1970. In any one year, HHS is equal to HH if the headship rates that year were the same as in 1940. If headship rates were lower in that year, HHS would be larger than HH and vice versa. The difference between HHS and HH thus is an aggregate indicator of change in headship rates.

Hickman estimated a regression equation that related net household formation (i.e. the change in HH) to the change in HHS and some economic determinants. These economic determinants consist of income, the price of housing, and the price of consumer goods in general. More specifically, these were formulated as:

YPDS: total disposable personal income in billions of dollars,
PYHR: implicit price deflator for rent of nonfarm dwellings, and
PCE: implicit price deflator for personal consumption expenditures.

The suitability of these as measures of income and price is discussed in chapter eight.

Hickman makes two discoveries of particular interest here. First, he finds that the price of housing (PYHR) had little effect on household formation. As one might expect, the rate of household formation declined as housing became more expensive. However, the long-run price elasticity of household formation was only about −0.01. Second, he finds that income had a more substantial impact. The long-run income elasticity of household formation was about 0.18. In other words, a 1 per cent rise in real income led to a 0.2 per cent rise in the number of households holding HHS constant.

A similar income elasticity for Canada would suggest a major role for the growing-affluence explanation. Between 1946 and 1981, Canadian real personal disposable income rose by about 150 per cent. At an income elasticity of 0.18, the long-run effect of this growing prosperity would have been about 27 per cent more households, holding constant population. The number of households actually increased by about 180 per cent over the same period. If two-thirds of this was attributable to a growing population, Hickman's estimate suggests that up to one-half of the remainder might be attributable to rising prosperity.

The British Experience

Hickman's analysis can be criticized on the basis of its aggregative nature. Overall, Hickman found an income elasticity of 0.18 and negligible price elasticity. However, these elasticities may well differ from one demographic group to the next. To what extent does an overall elasticity estimate hide important variations within the population?

Markandya (1983) takes a small step toward disaggregation in a study of household formation in Great Britain. In that study, headship rates are defined for the following groups of households:

FHALL: family (i.e. PNF) households,
NFHORA: other households (PNFP or NPNF) containing at least one person of retirement age (60 for women, 65 for men), and
NFHURA: other households (PNFP or NPNF) with no one of retirement age.

Using annual headship rate estimates for the period from 1961 to 1979, a separate regression is fitted for each group. The independent variables include the ratio of the housing component to the overall CPI (the author calls this RCOH), as well as distinct income measures for each type of household.

Interestingly, Markandya finds that RCOH had a positive impact on household formation among nonfamily units (i.e. NFHORA and NFHURA). In other words, a rise in the price of housing led to more household formation, not less, as reported in Hickman. Further, the long-run price elasticities are large: 0.5 for NFHORA and 1.4 for NFHURA. Only among family households was the price elasticity negative and insignificant. Markandya attributes the positive price elasticities in part to the role of capital gains. The rise in RCOH over this period partly reflected rapid increases in the price of owner-occupied housing. Households may have formed at an increasing rate when RCOH rose in order to capture these capital gains. Among families, the story is different, because they were traditionally formed households in any case. Household formation was sensitive to price, but largely only among nonfamily persons.

Markandya also finds that income elasticities are higher among nonfamily units. FHALL had a long-run income elasticity of about 0.2, the same as reported by Hickman. However, for NFHURA and NHFORA, elasticities of 1.2 and 0.6, respectively, are reported. Given the predominance of family households, these findings are not necessarily inconsistent with Hickman but could indicate a modestly higher income elasticity in the United Kingdom.

Markandya's results suggest that Hickman's income elasticity of 0.2 may be misleading. In addition to the fact that there were substantial differences among types of households, there was also a shift in the mix of households. As in Canada, the rising incidence of nonfamily households meant that the overall sensitivity of household formation to income or price changes was increasing. Hickman's 0.2 income elasticity represents an average over the 1922 to 1970 period. Probably, it was lower in the prewar years and rose through the postwar period.

The Canadian Experience

Smith's analysis for Canada is similar to Markandya's in that he disaggregates an overall household headship rate. He too distinguishes between family and nonfamily household formation but disaggregates further by age of head. For each year between 1962 and 1980, Smith estimates two headship rates (one for heading a family household, the other for heading any other type of unit) for each of four age groups. He then describes eight regression equations, one for each age-specific headship rate.

Smith includes as an independent variable the ratio of the real cost of housing (R) to income (Y). R is taken to be the CPI housing rent component. See chapter eight for a description of this measure. Y is total real per capita disposable income. Note that this income measure is not age-specific. Unlike Markandya's measures, it is an aggregate for all household types. Thus, the variable R/Y consists of the same set of annual data for all eight regressions.

Smith finds that R/Y had a significant negative effect on the headship rate for most age groups. In other words, a fall in the cost of renting relative to income increased the headship rate. The relationship was especially strong for nonfamily household headship. This is similar to Markandya's finding. Family household headship, in contrast, was significantly determined by R/Y only among the young. Smith finds that over age 35 family household headship rates were not substantially related to R/Y.

However, caution should be exercised in interpreting these findings. Over the period 1962–80, household headship rates rose substantially. At the same time, R/Y fell sharply with the rapid growth of real income and the sluggish increase in the CPI housing rent component. It should not be surprising, therefore, to find a strong statistical relationship. However, this need not imply causality. Headship rates may have risen over time for some other reason. The regression equation estimates merely reflect the fact that R/Y was simultaneously falling. If the temporal pattern of headship rates had involved ups and downs, and these had been matched by changes in R/Y, Smith's findings would be stronger. Much the same criticism can be levelled at Markandya. Hickman's analysis is somewhat more robust in this regard because his time series covered a sequence of economic booms and busts with corresponding changes in headship rates.

The remainder of this chapter considers some cross-sectional evidence for Canada. Cross-sectional evidence has the advantage that we can match individuals and families at different income levels to their living arrangement. After controlling for demographic variables that might shape living arrangement, we can isolate the distinct role of income. This allows us to get around the problem faced by these time series analyses.

CHOICE OF LIVING ARRANGEMENT

Let us define a family, or a nonfamily individual, to be a "unit." This chapter focuses on the living arrangements of units. Let us define two propensities. The first, $q[1]$, is the propensity to be a primary unit (i.e. to maintain a dwelling). Equivalently, a unit is said to be primary if the household head is a member of the unit. A lodger is defined to be any unit not maintaining a dwelling. Of course, for every lodger, there is a corresponding primary unit. The second propensity, $q[2]$, is the propensity for a primary unit to have no lodgers, i.e. to live alone. Figure 9 illustrates these definitions.

FIGURE 9

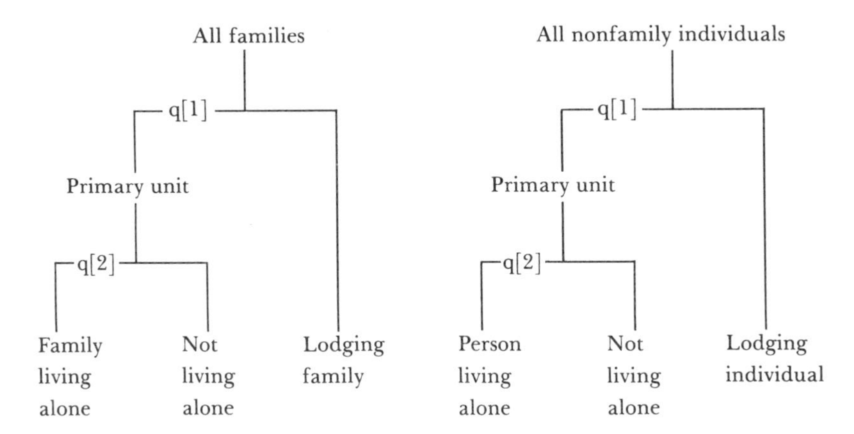

These two propensities provide a way of describing variations in living arrangement. However, they do not tell the entire story. Earlier, the family was pictured as an elastic unit because its definition requires coresidency. Since an individual's family status depends on whom he or she lives with, the definition of a unit itself is a function of living arrangement. For example, a young single adult might choose to remain in the parental home, thus being part of a family, or move out, likely becoming a nonfamily individual. The empirical work described later in this chapter does not consider the effect of changing living arrangements on family status. Rather, it focuses only on changes evidenced in $q[1]$ and $q[2]$.

Characterizing living arrangements in terms of $q[1]$ and $q[2]$, although clear in many cases, can be fuzzy. For most households - i.e. those consisting of either an individual or a family living alone - there is no ambiguity about the meaning of maintaining a dwelling. However, in cases where two or more units share a dwelling, the designation of headship and hence of the primary unit is not clear. Instance an elderly couple living with a son and daughter-in-law. In the 1971 Canadian Census, either the father or the son could be the household head. The corresponding couple was the primary unit, and the other couple a lodger. In such cases, the propensity to maintain a dwelling reflects social conventions as much as any substantive difference in living arrangement. However, this fuzziness need not arise in all cases of shared accommodation. For instance, where one unit pays rent to another coresiding unit (e.g. roomer paying rent to a landlord), the unit receiving the rent (i.e. the landlord) is unambiguously the primary unit.

Despite the fuzziness, these propensities are useful measures of living arrangement. Over the postwar period, it became more common for units of all types to maintain a dwelling and to live alone. Hence, $q[1]$ and $q[2]$ both increased. However, there is indirect evidence that the effects of income on $q[1]$ and $q[2]$ may have been different. Of the two, $q[1]$ appears to have been the more responsive. The number of families not maintaining a dwelling fell steadily from 1951 to 1971. It rose modestly after that in absolute terms but continued to decline as a proportion of all families. By 1981, young lone parents made up the only group of families wherein $q[1]$ was low, and these were families that typically had low incomes. Hence, $q[1]$ increased in step with rising prosperity, remaining low only among low-income groups. Over the postwar period, $q[2]$ also rose. However, available census data suggest that, at any one point in time, $q[2]$ was not strongly related to income. High-income units were not much less likely to take in a lodger than were their lower-income counterparts. In this sense, $q[2]$ may not have been as sensitive to changes in income.

What determines a unit's choice from among alternative living arrangements? We can think of this as an economic choice problem, i.e. in terms of consumer's income, preferences, and the prices of alternatives. For the moment, let us focus on the choice between living alone and shared accommodation (as either the primary or lodger unit). Commonly, shared accommodation is less expensive than living alone because of the consumption economies of shared accommodation. A dwelling usually has only one front door, or one furnace, regardless of the number of occupants. Increasing the number of occupants in a dwelling reduces the per capita shelter cost. In addition, shared accommodation can result in a more efficient use of labour. A retired grandmother, for example, may take care of the children during the day while the parents are working, obviating the expense of daycare. As another example, an otherwise unemployed person can take in lodgers, thereby becoming a gainfully employed housekeeper.

However, sharing accommodation imposes new costs as well. When units share accommodation, issues arise as to the purchase, care, and disposition of communal property. Roommates who want a sofa, for example, have to decide who is going to pay for it, the rules to be followed in using it, and what happens to the sofa if the household splits up. Congestion is another cost of shared accommodation. You might want to bake a cake at the same time as your roommate is using the oven for a roast.

Aside from the prices of different living arrangements is the issue of preferences. Even given identical prices for different choices, a unit may prefer one living arrangement to another. One aspect of preferences is the desire for privacy, however defined. To the extent that shared accommodation involves less privacy, it may be less preferred. However, for the young single, lodging may in fact mean more privacy than would be experienced in the parental home. In sharing accommodation, you may also find that your roommate's perceptions of acceptable or proper behaviour are not the same as yours, reducing one's preference for shared accommodation.

Risk aversion may also play a role in determining preferences. A lodger's rent provides an income to offset a drop from the unit's other sources of income. For example, a lodger's rent provides a supplementary income in the event that someone in the primary unit becomes unemployed. In this respect, a unit that is averse to the risk of a sudden drop in income might be more willing to take in a lodger than would a risk taker.

Let us now consider variations among the three types of units in terms of incomes, prices, and preferences. To what extent might their choices of living arrangement differ as a consequence of these considerations?

Husband-Wife Families

In the postwar period, husband-wife typically were better off than other kinds of units (see chapter eight for more detail about incomes). They were thus better able to afford separate accommodation. The less affluent among them - principally young and elderly couples - were more likely to lodge. Also, husband-wife families were better able to live alone than were other units, since each contained two adults, and possibly other family members, to help with household chores. Finally, husband-wife families were typically larger than other units. Given that relatively few dwellings contained more than three bedrooms, larger units typically did not have much leftover space in their dwelling with which to share accommodation.

Earlier, we suggested why husband-wife families became less likely, over the postwar period, to take in unrelated lodgers. In part, this was attributed to rising incomes. In part, it reflected an increase in the demand for privacy. Also, in part, it reflected the diminished value to the family of the rental income from lodging; the introduction of unemployment insurance and other welfare programs obviated the need for a second source of income. Also, the increasing incidence of working wives meant that less time was available for the additional housekeeping involved in having boarders.

Also of interest was the decline in lodging even among relatives. Husband-wife families in particular became less likely to have relatives living with them. In part, rising income made separate accommodation affordable to a family. In part, that same rising income allowed the family to support adult children and other relatives who otherwise could not have afforded separate accommodation, either directly via cash subsidies or indirectly through public programs such as subsidized student loans. Perhaps, also, there were changes in attitudes with regard to privacy. Shorter (1977, 205) argues that there was an increasing emphasis on social ties within the nuclear family, as opposed to ties with the rest of the community (including relatives): "The nuclear family is a state of mind rather than a particular kind of structure or set of household arrangements. It has little to do with whether the generations can live together or whether Aunt Mary stays in the spare bedroom . . . What really distinguishes the nuclear family . . . from other patterns of family life in Western society is a special sense of solidarity that separates the domestic unit from the surrounding community. Its members feel that they . . . enjoy a privileged emotional climate they must protect from outside intrusion, through privacy and isolation."

Few attempts have been made to measure the effect of income on the living arrangements of husband-wife families. One of these is described in Steele (1979, 108–11).[1] Using 1971 Census data for Toronto and Montreal, Steele found that higher-income families were more likely to live alone. However, she found also that living arrangements were not very sensitive to income, that there were only small differences in the propensity to live alone between low- and high-income units.

Lone-Parent Families

The living arrangements of lone parents differed substantially from those of husband-wife families. Most lone parents in the postwar period were women, and many had low incomes. On the basis of this alone, one should find different living arrangements. Further, many lone-parent families arose from marital disruptions (separation, divorce, or widowhood) that necessitated an immediate decision about where the family would live. Some lone parents, especially those who were younger and had few children, chose to move in parents who could provide moral, financial, or physical support.[2] Sweet (1972) argues that women with older (or more) children were themselves typically older and hence found it more difficult to move in with parents, were less in need of assistance in child rearing, had better jobs, and were less likely to consider their present marital status as temporary. Consequently, such women were more likely to maintain a dwelling. Some of them continued on in the family home, while receiving financial support from an estranged husband or an estate. Others sought new accommodation elsewhere. In these respects, the living arrangements of lone parents differed from those of a husband-wife family.

In looking at the effect of income on living arrangement, it is valuable to distinguish between lone parents by the ages of their children. Lone parents with older children might well have living arrangements like those of husband-wife units at a comparable income. However, lone parents with young children often found it difficult to earn an adequate income and raise children at the same time. Their living arrangement reflected the tradeoffs involved. For example, lone parents may want privacy just as much as husband-wife families do; however, only by sharing accommodation with someone who can help in child rearing might a lone parent be able to earn a good income. Because their job opportunities were more limited, lone parents living alone may have had lower incomes than those who did not. In contrast, husband-wife units living alone tended to have higher incomes than those who did not.

Of course, it is not necessary for lone parents to have someone around to help with child rearing. Daycare, for example, might be a substitute. A lone parent can leave small children at a daycare facility during workdays instead of having someone at home to look after them. Thus, the effect of the rise of daycare, particularly in the 1970s, on the living arrangements of lone parents should not be underestimated.

Sweet's study (1972) is one of the few to measure the effect of income on the propensity of estranged female lone parents to maintain a dwelling. His work distinguishes between earned income (i.e. from paid employment) and other income (e.g. welfare, interest, and support payments). After controlling for the woman's age, education, race, marital status, and the number and ages of children, Sweet finds neither earned nor other income to be an important determinant of household headship.[3]

Nonfamily Individuals

During the postwar period, the living arrangements of nonfamily individuals changed drastically. We have already noted the rapid increase in persons living alone. However, nonfamily individuals increasingly came to form other kinds of households as well. Glick (1976), in examining the living arrangements of 18-24-year-olds, found that the largest increase in U.S. household formation between 1960 and 1974 was among men living with unrelated persons. The number of such households grew by 461 per cent, compared to a 350 per cent increase in living alone.

What brought about the rise of the nonfamily household? In chapter four, some contributing factors are identified, including improved housing affordability; the declining supply of rooming facilities; improved home-making technology; the emergence of in-home support services for the elderly, handicapped, or ill; the growth in real incomes generally; and improvements in pension plans and other welfare schemes that helped raise the incomes of nonfamily individuals. Also important was the increased provision of subsidized housing, particularly for senior citizens. The rise of the nonfamily household was also a result of the postwar construction of apartment buildings, accommodation that is more suited to the needs of nonfamily individuals. Another important factor was increased geographic mobility. The postwar period was characterized by high volumes of long-distance migration by young adults, from one province to another and between rural areas, small towns, and larger cities. In many cases, young adults left parental homes behind when they moved. For migrants, living with parents or other relatives may simply not have been possible. Because of this, one might expect to find more nonfamily household formation among migrants than among nonmovers.

Michael et al (1980) wrote one of the few studies to look at the effect of income on the living arrangements of nonfamily persons. They looked at singles aged 25 to 34 and widows 65 or older. For each cohort, they calculated the proportion living alone in each U.S. state in 1970. Their study attempts to unravel the contribution of income to one-person household formation over time.[4] They conclude that the impact of income on the decision to live alone is nonlinear; there is an income range within which living arrangement is very sensitive to income. Around a 1970 income of $5,000, for example, they found an income elasticity of 1.0. Below or above this range, the propensity to live alone was found to be less sensitive to income. In regard to young singles, they conclude (45): "The intuitive interpretation implied by the S-shaped relationship between the propensity to live alone and income is that a threshold income was reached sometime in the 1940s after which further increases in income had a sizeable impact on the decision to live alone." They conclude that, between 1950 and 1976, the growth in real income alone accounted for over 70 per cent of the explained interstate variation in propensity to live alone for these two cohorts. Thus income growth was an important influence in the rise of the one-person household.

Was income growth as important as Michael et al's study suggests? It is difficult to generalize on this point because their study looks at just two cohorts and their data are aggregate (i.e. statewide averages). Aggregation can impart substantial biases to income elasticities. Chevan and Korson (1972) looked at the living arrangements of widowed persons using the 1960 U.S. Census. Their sample consisted of data for individuals, rather than statewide averages. They found that, although income was statistically significant in determining whether a widowed person lived alone, it accounted for only a small part of the variation in living arrangement. Harrison (1981, 35–50) is another individual-based study. Using data from the 1971 Census of Canada, he found that the decision to live alone was sensitive to income. Among nonfamily individuals under

TABLE 39
Living arrangements of nonfamily individuals by age and income group, Canada, 1971

		1970 income group			
	All incomes (%)	*Under $5,000 (%)*	*$5,000–$9,999 (%)*	*$10,000–$14,999 (%)*	*$15,000 or more (%)*
Under 35	100	100	100	100	100
Living alone	23	17	32	44	48
In nonfamily household*	40	40	41	36	32
In family household	37	43	27	21	19
35-54	100	100	100	100	100
Living alone	41	35	47	54	62
In nonfamily household	34	36	33	32	27
In family household	25	30	20	14	11
55 or older	100	100	100	100	100
Living alone	48	47	56	60	63
In nonfamily household	26	26	29	29	28
In family household	25	28	14	11	9

SOURCE: Data calculated from Harrison (1981, 38)
* Nonfamily household is any containing no nuclear family units.
† Family household contains at least one family unit.

35, for example, those with a 1970 income of $15,000 or more were three times as likely to live alone as those under $5,000. Among older individuals, the effect of income is somewhat smaller but still substantial. Some of his results are reproduced in Table 39. However, income was not found to be as important a determinant of living arrangement as Michael et al suggest. Many higher-income nonfamily individuals did not live alone, and many lower-income individuals did.

There is an additional problem. All these studies focus on one-person headship rates. For nonfamily individuals, this corresponds to the product q[1] and q[2]. Has income growth affected q[1] and q[2] differently?

DATA SOURCES AND ESTIMATION METHOD

Extensive data are needed to look at how living arrangement varies with income. Published census reports are simply not adequate. Instead, this chapter makes use of the census public use samples. As described in chapter five, these are 1 per cent random samples of census forms providing individual responses to selected questions. There are three samples: one of individuals, another of families, and a third of households. See chapter five for a more complete description of these samples. All three samples are used in this chapter.[5] To estimate the sensitivity of q[1] to income, we use the individual sample for nonfamily individuals and the family sample for family units. For q[2], we use the household sample for nonfamily individuals and the family sample for families. The 1971 public use samples are employed.

A unit's choice of living arrangement depends on its demographic characteristics. In part, demographic characteristics affect the prices of alternative living arrangements. In part, these characteristics also shape the unit's willingness to substitute between housing and other uses of the unit's income. To account for such variations, the two types of family units were subdivided by size, and all types of units were subdivided by age of head.

The census does not provide much information on the prices of alternative living arrangements. However, it is often argued that accommodation in general is more expensive in larger cities. The following analysis subdivides by place of residence: cities (urban areas of 30,000 population or more) v. other areas. This provides a crude method for looking at the effect of price on living arrangement.

From the public use samples, we know how many units were sampled within each group. We also know the total income of the family unit and the nonfamily individual's income. We know whether the unit lodged or maintained a dwelling. If the latter, we know also if anyone else was present in the household.

Just what is meant by income? In everyday language, income is often equated with cashflow, i.e. the total amount of money received by a consumer, regardless of source. More broadly, one might want to also include in kind, as well as cash, receipts. This is not, however, how income was defined in the 1971 Census. There, income included wages and salaries, commissions, bonuses, and tips; net income from self-employment, nonfarm business, or professional practice; net farm income; and receipts from family and youth allowances, government and private pension plans, other government income (unemployment insurance, veterans' pensions and allowances, and welfare were instanced), bond and deposit interest, dividends, other investment income (net rent revenue was instanced), and other income (alimony was instanced). Although lengthy, this list does not include other money receipts - such as money gifts from persons outside the consuming unit, inheritances, lump sum insurance payments, windfall gains, or capital gains - that might be considered part of income. In addition, it does not include other amounts - such as money receipts arising from the liquidation of assets or an increase in indebtedness - that also form part of a consumer's cashflow. Finally, it does not include noncash receipts.

The distinction between income and cashflow may seem dry and uninteresting. It should not be. As Brady (1958, 269), in focusing on private (between household) transfers, argues:

> Family relationships, by prescribing the responsibilities of individuals for other persons, govern the . . . distribution of income and . . . the allocation of consumer expenditures . . . A man's income may be distributed . . . to his wife and minor children sharing his home, to a son or daughter away at college, and to his elderly parents residing in another place. The commitments individuals make for the support of relatives determine the composition of consumer units . . . and the ways in which the needs of their members are satisfied. With sufficient income a man can contribute to the maintenance of his parents and of a grown son or daughter in separate living quarters but low incomes may impel the three generations to dwell in one household.

Much of the rapid postwar household formation occurred among low-income units, units that traditionally could not afford separate accommodation. Such units may well have experienced a substantial increase in cashflow, compared to income, during this period. In this sense, income may have become a less appropriate measure of the ability to afford such units.

Consider just two instances. One is the young individual or couple who receives economic support from parents, as in Brady's description above.[6] It has been argued that such economic support became increasingly important in the postwar period. With increasing prosperity, more parents were able to provide such assistance. In this way, young individuals and families found separate accommodation more affordable. A second instance is the young adult who receives a student loan while at college. The loan is not part of the student's income in census terms: in fact, it represents an increase in indebtedness. However, the student typically thinks of the loan money as just another cashflow receipt, to be used in defraying living expenses.

I suspect that the incidence and extent of such cashflow receipts increased markedly during the postwar period. For a critical group of low-income consumers, it made separate accommodation affordable for the first time, hence increasing the incidence of household headship. Because of this, it is important to be cautious in assessing the impact of income, in census terms, on household formation. Living arrangements may have been shaped more by cashflow than by income. At the same time, an almost complete absence of data on the cashflow receipts of households makes any empirical assessment difficult.

A logit model can be used to estimate the effect of income on living arrangement. Let us begin with a model for $q[1]$. Let Y be 0 if the unit is lodging and 1 if it is primary. Let W be the unit's income. The logit model asserts:

$$q[1] = z_1/(1 + z_1), \text{ where } z_1 = \exp(a_0 + a_1 W). \tag{20}$$

Above, a_0 and a_1 are constants to be estimated, and $q[1]$ is the probability that the unit will be primary. For a nonzero a_1, this yields a familiar S-shaped, or logistic, curve. We can think of the logit model as estimating the probability of forming a primary unit. Remember, though, that this probability, $q[1]$, is not observed directly. We observe only whether a unit is maintaining a dwelling (i.e. $Y = 1$ or $Y = 0$).

The parameter a_1 measures the sensitivity of $q[1]$ to income. If it is near zero, variations in income do not cause substantial changes in $q[1]$. A larger value means that living arrangement is more sensitive to income. A related measure of sensitivity is income elasticity. The two are related.

$$\text{Income elasticity} = a_1 W * (1 - q[1]). \tag{21}$$

Because of the shape of the logistic curve, income elasticity varies along its length. If a_1 is positive, income elasticity is highest at modest incomes and smaller at higher incomes. To make comparisons easier, all the income elasticities reported in this chapter are calculated at the same fixed income ($5,000).

A similar model is employed for $q[2]$. It takes the following form:

$$q[2] = z_2/(1 + z_2), \text{ where } z_2 = \exp(b_0 + b_1 W). \tag{22}$$

Again, b_0 and b_1 are constants to be estimated, and an income elasticity can be calculated. This elasticity measures the sensitivity of the decision to live alone to income among primary units.

Both sets of coefficients are estimated using a maximum likelihood criterion. In other words, the estimates are chosen so as to maximize the likelihood of getting the observed sample of $Y = 0$ and $Y = 1$ values given the units' incomes.

Ideally, one wants to study how the living arrangements change over time in response to a change in income. In order to do this, we would need to follow a sample of consumers over time: i.e. a panel study. However, there have been no large panel studies of consumers in Canada. Instead, we must rely on cross-sectional samples such as censuses that compare the living arrangements of different individuals at different incomes. Such data are used to estimate how an individual's living arrangement might change with an increase in income. However, there is a danger here because of the interdependence of living arrangement and income. Consider a young single adult attending university. She faces certain choices. On her income from summer work, she may be able to afford only a shared apartment. If, however, she takes a part-time job during the school year, she could live alone. If living alone is important to her, she might taking the part-time job to make that choice affordable. If it is not, she might prefer to use the extra time to pursue her studies. One person may prize privacy and choose the first alternative, another may have different priorities and choose the second. The point is that, for some consumers, income and living arrangement are interdependent choices.

In cross-sectional data, there tends to be a strong relationship between income and living arrangement. People who live alone tend to have higher incomes. However, this does not mean that poorer persons will be more likely to live alone as their incomes increase. In the above example, part of the reason the student might have a low income is because living alone is not important.

Thus empirical estimates of the impact of income on living arrangement, based on cross-sectional data, can overestimate the temporal effect of rising incomes. In this sense, the approach employed in this chapter may overstate the contribution to household formation of rising incomes over the postwar period. This should be kept in mind in evaluating the empirical findings.

INCOME AND THE LIVING ARRANGEMENTS OF HUSBAND-WIFE FAMILIES

The results from applying a logit model for $q[1]$ for husband-wife families are summarized in Table 40. Panel (a) contains results for families in cities of 30,000 population or more; panel (b) is for families in other areas. Each line in the table corresponds to a group of husband-wife families for whom a logit model for $q[1]$ was estimated. Because small samples can give unreliable estimates, results are not shown for group samples of fewer than 100 observations.

Consider the first row in panel (a). In the 1971 Census public use family sample, there were 1,007 two-person husband-wife families (i.e. no children present) in which the husband was 15-24 and the family resided in a city. The average 1970 total family income was $8,200. About 92 per cent of these families maintained a dwelling. In the logit model, a_1 is estimated to be 0.215. The asterisk indicates that this is at least twice the coefficient's asymptotic error, a measure of statistical significance. Using the estimated logit model, the expected probability of being a primary unit is 88 per cent at a family income of $5,000, and 95 per cent at $15,000, these being two points on the estimated logistic curve. Finally, the right-hand datum indicates that the income elasticity of $q[1]$ at an income of $5,000 was 13 per cent. This means that at a $5,000 income, there is a 0.13 per cent increase in expected $q[1]$ with a 1 per cent increase in income.

TABLE 40
Income and the propensity to maintain own dwelling by geographic area, husband-wife families, Canada, 1971

Size of family, age of husband	*Mean* $q[1]$* (%)	Mean w† ($000)	Sample size	$q[1]$ model estimates a_1‡	$q[1]$ w = 5 (%)	$q[1]$ w = 15 (%)	Elast.§ w = 5
(a) Cities							
Two persons							
15-24#	92	8.2	1,007	0.215*	88	95	0.13
25-34	94	11.2	1,596	0.144*	89	97	0.07
35-44	95	11.6	538	0.169*	90	98	0.08
45-54	98	10.6	997	0.190*	95	99	0.05
55-64	97	10.1	1,887	0.191*	95	99	0.05
65 or older	95	6.2	2,115	0.204*	95	99	0.05
Three persons							
15-24	93	7.0	563	0.148*	92	98	0.06
25-34	97	9.8	1,747	0.096*	95	98	0.02
35-44	98	11.2	810	0.145*	95	99	0.03
45-54	99	12.5	1,141	0.096	98	99	0.01
55-64	99	12.4	888	0.194	99	100	0.01
65 or older	97	10.7	334	0.082	96	98	0.02
Four persons							
15-24	94	6.5	143	0.056	94	96	0.02
25-34	98	9.7	1,922	0.106*	98	99	0.01
35-44	99	11.5	1,923	0.210*	98	100	0.02
45-54	100	13.4	1,373	0.115	99	100	0.01
55-64	99	13.9	519	0.228	96	100	0.04
65 or older	99	12.3	100	0.070	98	99	0.01
Five or more							
25-34	99	10.1	1,135	0.094	99	100	0.00
35-44	100	12.0	3,085	0.102	99	100	0.00
45-54	100	14.1	1,838	0.201*	100	100	0.00
(b) Other areas							
Two persons							
15-24	90	7.5	485	0.248*	87	99	0.16
25-34	93	10.0	675	0.175*	88	95	0.10
35-44	94	8.6	238	0.064	93	96	0.02
45-54	98	8.1	658	0.041	98	99	0.00
55-64	99	6.8	1,352	0.036	99	99	0.00
65 or older	96	4.7	1,933	0.147*	97	99	0.02
Three persons							
15-24	90	6.2	456	0.199*	89	98	0.11
25-34	96	8.2	957	0.193*	94	99	0.06
35-44	98	8.7	397	0.046	98	99	0.00
45-54	100	9.4	700	0.099	100	100	0.00
55-64	100	8.7	737	0.056	100	100	0.00
65 or older	99	7.0	358	0.169	99	100	0.01

(continued)

TABLE 40 (continued)

				$q[1]$ model estimates			
Size of family, age of husband	*Mean* $q[1]$* (%)	Mean w† ($000)	Sample size	a_1‡	$q[1]$ w = 5 (%)	$q[1]$ w = 15 (%)	Elast.§ w = 5
Four persons							
15-24	93	5.6	160	0.070	93	95	0.02
25-34	98	8.4	1,393	0.137	98	99	0.02
35-44	99	9.2	983	0.081	99	100	0.00
45-54	99	10.9	830	0.236	99	100	0.01
65 or older	99	8.2	112	0.269	99	100	0.01
Five or more							
25-34	99	7.8	1,275	0.127	99	100	0.01
35-44	99	9.1	2,737	0.116*	99	100	0.01
45-54	100	10.0	1,722	0.106	100	100	0.00
55-64	100	9.3	522	0.025	100	100	0.00

SOURCE: Computed from the 1971 Census public use sample (family file). See text for details.

Asterisk next to number indicates coefficient is at least twice its standard deviation.

* Proportion of families maintaining own dwelling

† Total family income for 1970

‡ a_1 is the coefficient of income ($000) in the logit model estimated using maximum likelihood

§ Income elasticity of propensity to maintain dwelling as estimated from the logit model

\# Rows omitted below where fewer than 100 observations or where all families maintain own dwelling

Look down the first column of Table 40. This gives the propensity to be a primary unit by geographic area, size of husband-wife family, or age of head, while controlling for the other two. For all groups considered, over 90 per cent of all husband-wife units maintained a dwelling. The incidence of lodging among husband-wife families was low. Further, this average $q[1]$ is higher the larger the family, after controlling for geographic area and age of head. Also, $q[1]$ tended to be lowest among the young, highest among the middle-aged (typically the 45-54 cohort), and slightly lower among the elderly. Finally, there was not much difference in the propensity to be a primary unit between cities and other areas.

However, these observations do not control for income. Above, it is argued that the price of housing tends to be higher in cities. Other things being equal, this should mean more shared accommodation there. However, if anything, $q[1]$ tends to be slightly higher in cities. Clearly, other things are not equal. In particular, incomes tend to be higher in cities, as evidenced in the second column of Table 40.

Are these income differences sufficiently large to account for the variation in $q[1]$? The remaining columns of Table 40 help us to answer this question. After fitting a logit model to each group, the estimated $q[1]$ at incomes of $5,000 and $15,000 was calculated. See the third and second from right columns of the table respectively. Note, from the "Mean W" column, that $5,000 was below average for a husband-wife family in 1971 and that $15,000 was above average. The difference in $q[1]$ between these two income levels indicates the variation in $q[1]$ that could be attributed to a substantial income difference. Let us first look at the estimated $q[1]$ at a $5,000 income. Even after controlling for income, larger families and middle-aged families were the most likely to be primary. Also, there are still no substantial geographic differentials. Husband-wife families in cities are no less likely than their counterparts elsewhere to maintain a dwelling. They may face higher housing prices, but, even at the same income, they are no less likely to be primary units. At the $15,000 income level, the estimated $q[1]$s are close to 1.0. There is little variation by size of family, age of head, or geographic location. Almost all higher-income families maintained a dwelling, regardless of their situation.

How important was income as a determinant of $q[1]$? The estimated income elasticities in the right-hand column of Table 40 provide useful additional detail. Young two-person husband-wife families had the highest elasticities. They show the sharpest distinction in living arrangements between lower- and higher-income units. Almost all older or larger families maintained a dwelling, regardless of income. Also, although income was important among young, smaller families, it did not account for much of the variation in $q[1]$ among modest-income families. Holding income constant, at $5,0000, there were still large variations in $q[1]$ with family size and age of head.

Let us now turn to $q[2]$. Logit model results are shown in Table 41.[7] Looking at the left-hand column, note that $q[2]$ was highest among young husband-wife families and lowest among the middle-aged. Also, $q[2]$ was higher among two-person units and lower for larger families. These are the opposite of the patterns found for $q[1]$. What they have in common is that neither shows much of a geographic differential. Given that young primary families are less affluent than their middle-aged counterparts (see the second column of Table 41), the first of these patterns is disturbing. Why aren't the wealthy more likely to live alone?

In part, this reflects the dominant rationale for sharing accommodation. Most husband-wife families live alone. Those with lodgers are principally extended families wherein the lodger is a grandparent, nephew, or other relative. Often taking in a lodger is a familial duty, rather than a commercial enterprise. During one's life span, the risks of such familial

TABLE 41
Income and the propensity to live alone by geographic area, primary husband-wife families, Canada, 1971

Size of family, age of husband	*Mean* $q[2]$* (%)	Mean w† ($000)	Sample size	b_1‡	$q[2]$ w = 5 (%)	$q[2]$ w = 15 (%)	Elast.§ w = 5
				$q[2]$ model estimates			
(a) Cities							
Two persons							
15-24‡	94	8.5	923	0.058	93	96	0.02
25-34	92	11.4	1,499	0.033	91	93	0.02
35-44	86	11.7	512	0.061*	81	85	0.06
45-54	85	10.7	972	0.018	83	86	0.02
55-64	85	10.3	1,826	0.022*	84	86	0.02
65 or older	89	6.4	2,007	0.016	89	91	0.01
Three persons							
15-24	91	7.1	523	0.043	90	94	0.02
25-34	90	9.8	1,690	0.033	89	92	0.02
35-44	84	11.3	790	0.047	80	87	0.05
45-54	84	12.5	1,129	0.009	83	84	0.01
55-64	86	12.4	883	0.070	79	88	0.07
65 or older	89	10.7	325	0.024	87	90	0.02
Four persons							
15-24	92	6.6	135	−0.005	92	92	−0.00
25-34	90	9.7	1,894	0.042*	89	92	0.02
35-44	88	11.6	1,903	0.028*	86	89	0.02
45-54	86	13.5	1,367	0.023*	84	87	0.02
55-64	88	14.0	512	0.028	85	88	0.02
Five or more							
25-34	90	10.1	1,127	0.039	88	92	0.02
35-44	90	12.0	3,075	0.018	89	90	0.01
45-54	89	14.1	1,836	0.002	89	89	0.00
55-64	89	15.4	423	0.041	85	90	0.03
(b) Other areas							
Two persons							
15-24	95	7.8	439	0.002	95	95	0.00
25-34	94	10.2	628	0.006	94	94	0.00
35-44	85	8.8	223	0.043	83	89	0.04
45-54	88	8.1	646	0.070*	86	93	0.05
55-64‡	88	6.8	1,335	0.023	88	90	0.01
65 or older	90	4.8	1,865	0.043	90	94	0.02
Three persons							
15-24	92	6.4	412	0.061	91	95	0.03
25-34	92	8.4	918	0.032	91	93	0.01
35-44	87	8.7	390	−0.028	88	85	−0.02
45-54	86	9.4	697	−0.005	86	86	−0.00
55-64	86	8.7	734	−0.010	87	86	−0.01
65 or older	90	7.1	355	0.024	90	92	0.01

(continued)

TABLE 41 (continued)

Size of family, age of husband	*Mean* $q[2]$* (%)	Mean w† ($000)	Sample size	b_1‡	$q[2]$ w = 5 (%)	$q[2]$ w = 15 (%)	Elast.§ w = 5
				$q[2]$ model estimates			
Four persons							
15-24	91	5.7	149	−0.054	91	86	−0.02
25-34	91	8.4	1,369	−0.016	92	90	−0.01
35-44	90	9.2	975	0.024	89	92	0.01
45-54	89	11.0	825	0.027	88	90	0.02
55-64	88	10.0	431	0.011	87	88	0.01
65 or older	86	8.3	111	0.119*	83	94	0.10
Five or more							
25-34	91	7.8	1,264	−0.003	91	91	−0.00
35-44	91	9.1	2,717	0.031*	90	93	0.02
45-54	89	10.0	1,718	0.022	88	90	0.01
55-64	87	9.3	521	−0.016	88	86	−0.01

SOURCE: As for Table 40

NOTES: As for Table 40, but in note ‡, for a_1 read b_1.

duty are highest in middle age. It is then that one's parents become elderly and less capable of maintaining their own home. It is also then that young adult nephews and nieces arrive in town to go to college or take a first job. Also, this is the age when one is best able to house a lodger. The family is more likely to be living in a detached house where an extra upstairs or basement bedroom can be more readily added. Also, with the home leaving of the first child, a middle-aged couple is more likely to have a spare bedroom.

The four right-hand columns indicate the insensitivity of $q[2]$ to income within each age group. Note the general lack of statistically significant results. The hypothesis that income does not affect the decision to live alone among primary units generally cannot be rejected. In fact, the slope coefficient is even negative (albeit insignificant) in some cases, suggesting that the propensity to live alone actually declined at higher incomes. For the most part, these results do not support the hypothesis that affluence accounted for the rise of the family living alone.

As additional evidence of the weak relationship between $q[2]$ and income, consider the low elasticities in the right-hand column of the table. These $q[2]$ elasticities are much smaller than many of the $q[1]$ elasticities discussed above. Overall, these results show that, at least in 1971, income was an important determinant only of $q[1]$, the propensity to maintain a dwelling, and only for young, small husband-wife families.

The propensity for a primary family to live alone was not sensitive to income.

INCOME AND THE LIVING ARRANGEMENTS OF LONE-PARENT FAMILIES

As noted earlier, almost all lone parents were women. With so few male lone farents, it is not possible to undertake a separate statistical analysis, even with the relatively large census public use samples. The following discussion is restricted to female lone parents.

As with other groups, female lone parents prospered in absolute terms in postwar Canada. However, their incomes continued to be low relative to other groups in society. A comparison of Tables 40 and 42 confirms that average family incomes of lone parents were considerably below their husband-wife peers.[8]

In certain respects, lone-parent and husband-wife families were similar in their propensity to lodge. There was little difference in the extent of lodging among lone parents between cities and other areas. Also, larger lone-parent families were more likely to be primary units than were smaller families.

However, unlike husband-wife units, the propensity to lodge showed a breakpoint about age 45. Lone parents over 45 were roughly as likely as husband-wife units to lodge. Under age 45, this was a more common living arrangement. As with husband-wife units, lodging was most prevalent among young and small lone-parent families. The difference is that lodging was more prevalent here than among young husband-wife units. Lodging was especially common among young, small lone-parent families living in cities.

The application of a logit model for $q[1]$ for lone-parent families is summarized in Table 42. To maintain comparability with Table 40, the estimated $q[1]$ at incomes of $5,000 and $15,000 is shown, although both were relatively high incomes for a lone-parent family. Note, in general, the lack of asterisks in Table 42. Although the a_1 coefficients are not, in general, numerically smaller than those in Table 40, relatively few are statistically significant. This reflects the smaller sample sizes for lone-parent families. The much larger samples of husband-wife units make even a weak numerical relationship statistically significant. At the same time, some of the estimated elasticities are quite large. The sensitivity of living arrangement to income was greatest among young, small lone-parent families. For example, two-person lone-parent families living in cities and headed by a woman under 25 had an elasticity of 0.32 at an income of $5,000, a much higher value than anything found for husband-wife families.

TABLE 42
Income and the propensity to maintain own dwelling by geographic area, lone-parent families headed by a woman, Canada, 1971

				q[1] model estimates			
Size of family, age of husband	*Mean* q[1]* (%)	Mean w† ($000)	Sample size	a_1‡	q[1] w = 5 (%)	q[1] w = 15 (%)	Elast.§ w = 5
(a) Cities							
Two persons							
15-24‡	61	2.1	127	0.352	82	99	0.32
25-34	75	3.3	146	0.199*	82	97	0.18
35-44	81	4.4	143	0.061	82	89	0.06
45-54	93	5.4	247	0.089	93	97	0.03
55-64	94	6.4	227	0.297*	95	100	0.07
65 or older	94	7.6	284	0.210*	83	99	0.07
Three persons							
25-34	88	3.6	143	0.127	90	97	0.06
35-44	90	4.9	147	0.295	94	100	0.09
45-54	96	7.6	169	-0.018	96	95	-0.00
Four or more							
25-34	93	3.0	137	-0.110	92	80	-0.04
35-44	97	5.1	226	0.319	98	100	0.03
45-54	98	7.6	164	0.582	99	100	0.03
(b) Other areas							
Two persons							
45-54	90	4.7	101	0.111	91	97	0.05
55-64	97	4.6	146	0.607*	99	100	0.03
65 or older	95	4.5	186	0.250	96	100	0.04
Four or more							
35-44	94	4.2	125	0.406*	97	100	0.06

SOURCE: As for Table 40
NOTES: As for Table 40

Let us now turn to $q[2]$, the propensity for primary units to live alone. The left-hand column of Table 43 shows average $q[2]$ values for each group of lone parents. As with husband-wife families, this propensity is somewhat lower among the middle-aged and higher among the young and the elderly. Again, this may reflect the dominance of extended families in the lodging phenomenon. The propensity to live alone was much lower among lone parents than among husband-wife families. In part, the lower incomes of lone parents made separate housing less affordable. However, this is not the whole picture. The absence of statistically significant results is also noteworthy. Nowhere is income a significant determinant of living alone among primary lone-parent families. Also of interest are the widespread negative b_1 coefficients, which suggest that more affluent lone parents were less likely to live alone, i.e. more likely

TABLE 43
Income and the propensity to live alone by geographic area, lone-parent primary families headed by a woman, Canada, 1971

				q[2] model estimates			
Size of family, age of husband	*Mean* q[2]* (%)	Mean w† ($000)	Sample size	b_1‡	q[1] w = 5 (%)	q[2] w = 15 (%)	Elast.§ w = 5
(a) Cities							
Two persons							
25-34‡	80	3.6	109	−0.099	78	57	−0.11
35-44	82	4.5	116	0.044	82	88	0.04
45-54	72	5.5	230	0.009	72	74	0.01
55-64	80	6.6	214	0.015	80	82	0.01
65 or older	79	7.7	268	−0.049	82	73	−0.05
Three persons							
25-34	84	3.7	126	−0.011	84	82	−0.01
35-44	80	5.1	133	−0.021	80	76	−0.02
45-54	86	7.5	162	−0.019	86	84	−0.01
Four or more							
25-34	86	3.0	128	−0.066	84	74	−0.05
35-44	80	5.2	220	−0.018	80	77	−0.02
45-54	80	7.7	161	0.091	76	89	0.11
(b) Other areas							
Two persons							
55-64	78	4.8	141	0.038	78	84	0.04
65 or older	82	4.6	177	0.064	83	90	0.05
Four or more							
35-44	81	4.4	117	0.023	81	85	0.02

SOURCE: As for Table 40
NOTES: As for Table 40, but in note ‡, for a_1 read b_1.

to take in lodgers. One can shrug off these cases by arguing that the results are not statistically significant. However, the number of negative results suggests a pattern, albeit weak. Low-income lone parents were, if anything, more likely to live alone than their higher-income peers.

Why? Given that separate accommodation is more expensive, should not lower-income families be more likely to share a dwelling? Not necessarily. In most cases, the income of a lone parent is constrained by her ability to do paid work. The responsibilities of child bearing may limit the extent to which she can work outside the home. Having a lodger (e.g. a relative) who can help with child rearing makes it possible to spend more time at a paid job, thus increasing her income. Thus, some lone parents have higher incomes specifically because they share their dwelling. Lone-parent families that live alone are typically less able or willing to seek paid employment and thus tend to have a lower income.

This argument applies mainly to lone parents with young children. As children age, the responsibilities of child rearing subside, and paid employment outside the home becomes more feasible. This should obviate the connection between living arrangement and income. However, Table 43 indicates some negative (albeit small) income elasticities even among older lone parents. This might be because of the impact of child rearing on lifetime earnings. A lone parent who had only a part-time paid job when her children were young may find it difficult in later years to advance her career. Her peer who had had a full-time job would have acquired more experience, knowledge, and seniority and had more opportunities for advancement. Thus, lone parents who took in a lodger in order to work outside the home would likely have a higher income, even in their post-child rearing years. However, having experienced shared accommodation, and perhaps being more amenable to lodging by nature, these same lone arents would be more likely to have lodgers even after the needs of child rearing had subsided. In this respect, higher-income lone parents, even with older children, would still be more likely to have lodgers than would their lower-income counterparts.

There is a danger in emphasizing such arguments, though, because they are based on weak empirical relationships. In general, lone parents at higher incomes are only modestly more likely to share accommodation than are their lower-income peers.

INCOME AND THE LIVING ARRANGEMENTS OF NONFAMILY INDIVIDUALS

Finally, let us consider the effect of income on the living arrangements of nonfamily individuals. In Table 44 are shown the propensities to be a household head (i.e. $q[1]$) for nonfamily individuals grouped by age, sex, and location. Nonfamily individuals were much less likely to maintain a dwelling than were husband-wife or lone-parent families. Nowhere does $q[1]$ rise much above 60 per cent in Table 44. In contrast, $q[1]$ averaged well over 90 per cent for almost all husband-wife and lone-parent family groups. Why were nonfamily individuals less likely to be household heads?

Part of the explanation lies in lower incomes. As seen in the second column of Table 44, the typical nonfamily individual was not well off. For example, young men under 25 had an average income of about $3,000 in 1970. In contrast, young husband-wife couples (without children) averaged over $8,000.[9] To a certain extent, this may explain the nonfamily individual's lower propensity to maintain a dwelling. However, a comparison with the incomes of lone-parent families belies this argu-

TABLE 44
Income and the propensity to maintain own dwelling by geographic area, nonfamily individuals, Canada, 1971

	Mean $q[1]$* (%)	Mean w† ($000)	Sample size	$\dot{q}[1]$ model estimates			
				a_1‡	$q[1]$ w = 5 (%)	$q[1]$ w = 15 (%)	Elast.§ w = 5
(a) Cities							
Men							
15-24#	26	3.0	1,833	0.169*	31	71	0.58
25-34	50	6.1	1,527	0.102*	48	72	0.27
35-44	52	7.0	850	0.089*	48	70	0.23
45-54	56	5.9	740	0.087*	54	74	0.20
55-64	56	5.7	647	0.045*	55	66	0.10
65 or older	53	3.9	981	0.094*	56	77	0.21
Women							
15-24	30	2.4	1,865	0.308*	47	95	0.81
25-34	50	4.5	1,059	0.249*	54	93	0.58
35-44	52	4.7	601	0.200*	54	90	0.46
45-54	61	4.5	904	0.207*	66	94	0.36
55-64	62	3.9	1,413	0.163*	67	91	0.27
65 or older	61	2.8	3,026	0.253*	75	97	0.32
(b) Other areas							
Men							
15-24	14	2.2	1,016	0.205*	21	67	0.81
25-34	38	5.6	502	0.083*	37	57	0.26
35-44	53	5.1	392	0.041	53	63	0.10
45-54	59	4.4	520	0.062*	60	74	0.12
55-64	65	3.8	569	0.049*	67	77	0.08
65 or older	62	2.5	958	0.118*	69	88	0.18
Women							
15-24	14	1.4	780	0.479*	40	99	1.43
25-34	24	3.3	234	0.259*	30	85	0.91
35-44	33	3.7	168	0.153*	37	73	0.48
45-54	52	3.0	344	0.160*	60	88	0.32
55-64	64	2.7	668	0.162*	73	93	0.22
65 or older	64	2.2	1,738	0.121*	72	90	0.17

SOURCE: Computed from the 1971 Census public use sample (individual file). See text for details.
NOTES: As for Table 40, but here w is individual income.

ment. Most lone parents (especially those with young children) had lower incomes than did nonfamily women of the same age and yet were more likely to maintain a dwelling. Factors other than income were important in shaping living arrangement.

From the elasticity estimates in the right-hand column of Table 44, income was evidently an important determinant of $q[1]$. The elasticity estimates range as high as 1.43, several times larger than anything encountered for families. The elasticities were low among the middle-aged but even here are large compared to anything estimated for husband-wife units. Also, note that these results are, for the most part, significant. In these two senses, income was an important determinant of the propensity to maintain a dwelling.

Consider now $q[2]$: the propensity for primary nonfamily individuals to live alone. See Table 45. In general, primary nonfamily individuals were much less likely to live alone than were their husband-wife or lone-parent family peers. Also, unlike family units, these individuals were increasingly likely to live alone with increasing age. There was no tendency for $q[2]$ to dip in the middle-aged groups. Whatever led middle-aged families to be more likely to take in a lodger did not apply to nonfamily individuals. They were not more likely to take in a lodger at that stage in their life.

The income elasticities for $q[2]$ are also of interest. As with family units, these elasticities are all small. Some are even negative. Very few are statistically significant. In general, they imply that the decision to live alone among primary nonfamily individuals was not sensitive to income.

However, this does not mean that growing prosperity was unimportant in shaping the rise of the one-person household. Remember that most nonfamily individuals who maintained a dwelling also lived alone. We have already seen that $q[1]$, the propensity to maintain a dwelling, was very sensitive to income. In that sense, income was an important contributor to the rise of the one-person household. However, among nonfamily persons maintaining a dwelling, the decision to take in a lodger was relatively insensitive to income. Most nonfamily household heads lived alone. There was no tendency for lower-income household heads to be more likely to share accommodation.

Before leaving this section, it is useful to contrast these results with those of Michael et al (1980). Different measurements, models, and underlying concepts make exact comparisons impossible. Nonetheless, the scale of the differences makes even crude comparisons valuable. In Table 44, the income elasticity estimates for $q[1]$ range from 0.26 to 0.91 for nonfamily persons aged 25-34. For women over 64, the estimates range only from 0.17 to 0.32. Both sets of figures are below the 1.0 value cited by Michael et al for 25-34-year-old singles and elderly widows. For $q[2]$, the estimated elasticities are still smaller. Finally, note that the youngest cohorts have the highest elasticities; all other cohorts are smaller. Based on these crude comparisons, Michael et al's estimates of the effect of income on household formation appear too high. This may be because the American experience was different. It may also reflect an aggregation bias arising from their use of statewide averages, rather than

TABLE 45
Income and the propensity to live alone by geographic area, primary nonfamily individuals, Canada, 1971

	Mean $q[2]$* (%)	Sample size	$q[2]$ model estimates b_1‡	$q[2]$ w = 5 (%)	$q[2]$ w = 15 (%)	Elast.§ w = 5
(a) Cities						
Men						
15-24#	50	519	0.032	51	59	0.08
25-34	62	683	0.018	61	65	0.04
35-44	69	466	0.044*	67	76	0.07
45-54	71	425	0.029	69	75	0.04
55-64	72	402	0.008	72	74	0.01
65 or older	72	512	−0.012	72	70	−0.02
Women						
15-24	55	525	0.010	55	58	0.02
25-34	68	481	0.036	68	75	0.06
35-44	75	346	0.029	74	80	0.04
45-54	72	53	−0.055*	73	61	−0.07
55-64	75	953	0.000	75	75	0.00
65 or older	78	1,770	−0.038*	77	70	−0.04
(b) Other areas						
Men						
15-24	60	152	−0.014	60	57	−0.03
25-34	63	223	0.063	61	75	0.12
34-44	64	201	0.042	63	72	0.08
45-54	64	301	0.003	64	64	0.01
55-64	73	359	−0.016	73	70	−0.02
65 or older	78	577	0.013	78	80	0.01
Women						
15-24	52	101	0.020	52	58	0.05
45-54	63	188	−0.033	63	55	−0.06
55-64	73	417	0.062	75	85	0.08
65 or older	82	1,106	−0.010	82	80	−0.01

SOURCE: Computed from the 1971 Census public use sample (household file). See text for details.
NOTES: As for Table 40, but here w is individual income, and in note ‡, for a_1 read b_1.

individual data. Further, the disaggregation into $q[1]$ and $q[2]$ elasticities emphasizes an important point. Nonfamily individuals may have used their rising incomes to purchase household headship, not necessarily to live alone. It is true that most nonfamily persons maintaining a dwelling lived alone. However, income was not important in determining whether they lived alone or had a partner.

INCREASING AFFLUENCE AND HOUSEHOLD FORMATION

Tables 40 through 45 present a wealth of data. Having briefly considered each alone, let us now compare them side by side. The basic question is still the same. Just how important was increasing affluence in the formation of households?

One way to address this question is to compare living arrangement at different income levels. In the above tables, we have estimated $q[1]$ from the respective logit models, each at incomes of $5,000 and $15,000. For husband-wife families, there is little difference in $q[1]$ between these two income levels, except for young, small families. Income was not an important determinant. For lone-parent families, the differences are somewhat larger, especially among younger women. Here, the growth in income may have had a larger role to play in the formation of new households. It is among nonfamily individuals, though, that the largest differences in $q[1]$ between low and high incomes are found. These differences are substantial, regardless of age, sex, and geographic location. Household formation among nonfamily persons is sensitive to income.

However, $q[1]$ measures only one aspect of household formation. It describes the propensity for a unit to maintain a dwelling. Because it looks at the decision to live alone among primary units, $q[2]$ provides additional information. For both husband-wife families and nonfamily individuals, there was little difference in q[2] between lower and higher incomes. For lone-parent families, the differences were again small, and sometimes negative. This does not mean that, as its income increased, a lone-parent family preferred to take in a lodger. Rather, it means that lone parents who wanted a higher income found it necessary to take on a lodger to help with child rearing while the mother worked outside the home. Overall, income was not an important determinant of $q[2]$.

Another way of assessing the importance of income is to look at the contributions of other factors. Reconsider $q[1]$. We can hold income constant in Tables 40, 42, and 44 by looking at the $q[1]$ model estimate at the $5,000 income level. Looking down this column, note that the propensity to maintain a dwelling increased with age, peaking among the middle-aged for families and continuing to rise through old age for nonfamily persons. In general, these age variations reflect the change in mean $q[1]$ in the left-hand column. Although there are specific groups where a low income is associated with a higher incidence of lodging, variations in mean $q[1]$ are related more to age than to income differences. For example, only about 31 per cent of 15–24-year-old nonfamily men, living in cities and with an income of $5,000, were maintaining a dwelling, compared with almost 48 per cent of men 25–34 at the same income.

Something other than just income was important in determining living arrangement.

The evidence also suggests that $q[1]$ was sensitive to family size, even after controlling for income and age of head. Take the case of 25-34-year-old husband-wife families in cities. Among two-person families at $5,000 income, 89 per cent maintained a dwelling. This increased with size to 99 per cent of families of five or more. Family size was thus also an important determinant of $q[1]$.

Finally, geographic location was of mixed importance. It appears to have been unimportant in shaping $q[1]$ for either husband-wife or lone-parent families. It was, however, more important for nonfamily individuals. In general, younger nonfamily individuals living in cities were more likely to maintain a dwelling than their counterparts elsewhere. This difference disappeared among older women and was reversed for older men. With the latter exception, these results are surprising. City residents normally face higher shelter costs; thus they should be less likely to maintain a dwelling than their counterparts elsewhere. These findings suggest the opposite, at least among young adults. Clearly, some other explanation must be sought.

Overall, income growth appears to have been just one of several factors that shape the decision to maintain a dwelling. Family status and size, age, and geographic location were also important. In some cases, these latter factors appear to have been at least as important as income. This suggests that postwar changes in marriage, fertility, age distribution, and geographic location may have played important roles in reshaping living arrangements.

This conclusion is reinforced when we turn our attention to $q[2]$. Reconsider the $q[2]$ model estimates (at $5,000 income) in Tables 41, 43, and 45. In general, the decision to live alone among primary units was not sensitive to income. Among families, the drop in $q[2]$ in the middle-aged groups runs counter to the higher incomes found here, further evidence that income was not an important determinant. Rather, stage in the family life cycle and the availability of a spare room might be better predictors. For the most part, these results are consistent with those of previous studies. For family units, there simply was not a strong relationship between income and living arrangement. The present results go further, by pointing out that the decision to live alone, $q[2]$, was especially insensitive to income. Further, only among nonfamily individuals was there evidence of a large income elasticity for $q[1]$.

Michael et al (1980) argue that the relationship between income and living arrangement for nonfamily individuals is nonlinear. They think that after the 1940s, the income of the typical nonfamily person reached a level at which the choice of living arrangement became quite sensitive.

They argue that postwar increases in income could account for the sudden, and unprecedented, surge in headship rates among nonfamily individuals. The results in this chapter provide some support for that argument. However, it is not evident that income growth could have accounted for all, or even a substantial portion, of the increase in headship rates.

At the same time, it is clear that income growth was important in shaping the decline of shared accommodation. It was not the rise in family incomes, though, that brought this about. Instead, it was the increasing incomes of nonfamily individuals. It was nonfamily individuals who used their rising incomes to purchase separate accommodation, more than families using theirs to live alone.

This brings us back once again to the concept of privacy. It has been argued that postwar families and individuals used their growing affluence to purchase the privacy of living alone. The evidence considered here does not suggest this. Rather, families and individuals tended to use their affluence to obtain a living arrangement that would give them control over their environment: i.e. household headship. To be sure, this may also have often meant living alone. However, the objective commonly was to maintain one's own dwelling, rather than simply to live alone.

CHAPTER SEVEN

The Postwar Housing Stock

Patterns of population growth, family formation, and household formation are important in any discussion of housing demand. However, they do not tell the entire story. The postwar demand for housing in Canada was a demand for both more housing and different kinds of housing. Aggregate growth in population, families, and households helps to explain why more housing units were demanded; however, it does not readily explain changes in the quality and amount of housing demanded.

This chapter describes how the housing stock changed in postwar Canada. At the outset, available data are discussed and their limitations illustrated. Censal estimates of aggregate housing stock growth are presented, and changes in structural type (e.g the great apartment boom of the late 1960s and early 1970s) are considered. This is followed by a discussion of tenure shifts: from owner occupancy to renting during the 1940s, a reversal in the 1960s, and the re-emergence of owner occupancy in the late 1970s. In the next part of the chapter, the changing size distribution of dwellings (in terms of rooms) is described and related to the changing size distribution of households (in terms of persons). Finally, postwar changes in different quality attributes of the housing stock are assessed; these include age of dwelling, dwellings in need of major repair, plumbing and toilet facilities, electrification, heating system employed, and exterior finish.

As in earlier chapters, the primary data sources are published census reports between 1941 and 1981. However, the 1956 Census did not cover dwelling characteristics, and the 1966 and 1976 censuses inquired only about tenure and dwelling type. Therefore much of the ensuing analysis is based strictly on the decennial censuses. Further, later housing data, other than dwelling type and tenure, are derived from the population subsample that responded to the long (2B) census questionnaire. In 1961 and 1981, the 2B form was given to just 20 per cent of households. In

1971, a one-third sample was used. In the 1941 and 1951 censuses, a separate housing questionnaire was administered to 10 and 20 per cent of the population, respectively. The data presented in published census reports are counts for all of Canada estimated from these various samples. This introduces an additional element of sampling variability (albeit small because of the large sample sizes involved) that must be considered when comparing different census findings.

ON MEASURING CHANGE IN THE STOCK

That no systematic records have been kept limits our ability to comment on how the housing stock changed in postwar Canada. To illustrate this problem, let us first discuss a satisfactory system of "housing accounts" data. Then we will consider the adequacy of existing housing data.

The purpose of a system of housing accounts is to identify how housing stock changes over time. At its most rudimentary, this system describes components of change in the aggregate stock. Consider the following example. The 1976 Census enumerated 7.567 million dwellings in Canada. See Table 46. This figure includes dwellings that are unoccupied or occupied by temporary or foreign residents but excludes seasonal or marginal dwellings except those occupied by "usual" residents. In the 1981 Census, the corresponding figure was 8.785 million dwellings. Between 1976 and 1981, the stock of dwellings increased by 1.217 million units. This net addition was effected partly by construction of new buildings and partly by conversion of older dwelling units into multiple units. At the same time, demolition, deconversion, conversion into nonresidential use, and abandonment reduced the net addition. Each of these can be thought of as a component of housing stock change. For some of these components, however, data are scarce or nonexistent.

Typically, the largest component is new dwelling unit completions. Divic (1982, 18) estimates that new completions totalled 1.112 million dwellings between 1 June 1976 and 1 June 1981. This is the component of housing stock change for which data are most readily available. Data on demolitions, conversions, and other components of housing stock change are less satisfactory. For example, Divic (1982, 18) estimates that there were 59,000 demolitions and 10,000 net conversions between 1976 and 1981. These estimates are for calendar years, however, not the June-to-June "census years" on which housing unit completions are based. Even allowing a reasonable adjustment to get to a census-year basis, these dwelling stock change figures do not exactly correspond. The components data suggest a change in the housing stock of $1.112 - 0.059 + 0.010 = 1.063$ million dwellings. The increase in the census

TABLE 46
Dwellings by characteristics, Canada, 1976 and 1981

	1976	*1981*
Total dwellings*	7,567,345	8,784,620
Private dwellings	7,550,895	8,756,675
Occupied by usual residents	7,166,095	8,281,530
Regular dwellings	†	8,273,985
Seasonal/marginal dwellings	†	7,545
Occupied by temporary/foreign residents	18,950	55,995
Unoccupied‡	365,855	419,155
Available	207,135	
For sale	71,985	
For rent	135,150	
Unavailable	158,720	
Collective dwellings	16,445	27,945

SOURCE: Taken from published reports of the 1976 and 1981 censuses of Canada

* The reader should be aware of the various kinds of errors that arise in enumerating a housing stock. Some of the more important of these are described in *1981 Census of Canada Unoccupied Private Dwellings: Limitations to the Data*. Cat. 93-X-901. Ottawa: Statistics Canada, 1983. The 1976 Census undercounted the number of occupied dwellings by about 2 per cent. An estimate for the 1981 Census is unavailable. In 1976, about 0.4 per cent of occupied private dwellings were misclassified as "unoccupied" in 1976; a correction was applied in 1981. In 1976, about 1.7 per cent of unoccupied dwellings were missed, and a similar percentage in 1981. In 1981, about 1.4 per cent of unoccupied dwellings were enumerated twice. These dwellings were enumerated as vacant. Later, they were found to be occupied, and a second enumeration was entered without removing the first. No estimate is available for 1976. In 1981, there was a major overcounting of unoccupied dwellings. An overcoverage rate of about 20 per cent is estimated because of the enumeration of structures under construction or renovation or seasonal dwellings. 1976 counts likely contain a similar level of overcounting.

† Data not available

‡ Excludes unoccupied seasonal/marginal dwellings

stock of dwellings, as noted above, was 1.217 million. This is a difference of 1.217 − 1.063 = 0.154 million, or just under 15 per cent of the components of change estimate.

What happened here? In a system of housing accounts, an observed change in the dwelling stock should exactly equal the sum of the components. In this case, though, it does not. There are several possible explanations for this discrepancy. One is that a statistical discrepancy is inevitable because the two kinds of data being compared – census counts of dwellings on the one hand, and housing completions, conversions, and demolitions on the other – are collected independently. The data collection system does not ensure that such counts conform. A system of housing accounts would require an ongoing inventory of dwelling units, tracing the creation of new units and the loss of older ones. At any point in time, we would thus know how many dwelling units are in the housing stock and how each one came to exist: e.g. a new unit completion or conversion. There would be an exact correspondence between housing stock change and the sum of the components.

A related explanation is that counts of conversions, demolitions, and abandonments are suspect. Typically, these changes are measured only in conjunction with some statutory requirement. For example, many local governments require a permit to demolish a building. These permit applications can be used to estimate the number of demolitions of dwelling units. However, even with relatively good permit data, there are problems. A detached house, for example, which had been flatted into two or more dwellings, may have its demolition recorded incorrectly as the loss of just one dwelling unit. The problem is more severe in the case of conversions. In estimating these, one usually must rely on information drawn from building permits. Most local governments require a building permit for new construction or major renovations, extensions, and repairs. From the information in such permits, it is sometimes possible to identify when an older dwelling unit has been converted into two or more, a dwelling has been converted to nonresidential use or vice versa, or when two or more older dwellings have been combined into a single unit. However, conversions can also occur without the issuance of a permit; in some cases, this is done surreptitiously, while, in others, the amount of construction work involved is too small to require a permit.

Little information is available on the overall extent of conversions in the housing stock. In one of the few empirical studies of conversions, Morrison (1978) used a sample of detached dwelling properties in Metropolitan Toronto. The building on each property sampled was constructed prior to 1947 and still standing between 1957 and 1973. Morrison recorded the number of dwelling units found in the building (its "size category" in his terms) in each property at any time between 1961 and 1971. Morrison (1978, 129) concludes: "From 1961 to 1971, over one third of all sampled properties in the inner urban area shifted from supplying dwelling units in one size category to that in another. Nearly twenty-five percent of properties had one or more additional dwelling units produced from them and just over ten percent of properties sampled had one or more dwellings removed." Morrison's study design focused on a geographic location and dwelling type where conversion activity was widespread. Nonetheless, his work indicates that these components of housing stock change may be considerably more important numerically than has been widely recognized. The lack of adequate data on these components is thus troublesome.[1]

To this point, we have considered only the aggregate stock. Typically, however, we are interested in the components of change for particular kinds of dwellings: e.g by tenure, type, size, or quality level. For instance, we might want to know the extent to which inferior housing was eliminated by upgrading through renovation or by demolition. Or we might be interested in the extent to which the stock of rental housing was increased by new construction or by conversion from owner occupancy. To examine such issues, we need a more elaborate system of housing accounts, which traces the components of change separately for each kind of housing (including data on transitions between these different kinds).

However, no such system of housing accounts existed in Canada at any time during the period of study. Data on conversions, in particular, are very scarce. It is thus difficult to describe systematically how Canada's housing stock changed. For the most part, we must rely on two sources of data: quinquennial census descriptions, "snapshots" of the housing stock as it then existed; and annual counts of dwelling unit completions.

Census reports detail the housing stock as it then exists. These include, depending on the census, information on tenure, dwelling type, size, condition, period and mode of construction, and geographic locale; rents, house prices, and other shelter costs; and counts of basic facilities and equipment such as method of water-supply, heating equipment, and availability of electricity. Some of these data can be used as indicators of quality. However, remember that these data are "snapshots." They do not tell us how the dwelling units have changed since the previous census. Also, these data typically cover only private, occupied dwellings. As seen in Table 46, private, occupied dwellings are not exactly the same as the total stock of dwellings.

Data on dwelling unit completions are derived from Canada Mortgage and Housing Corporation's Starts and Completions Survey (SCS). This was a complete monthly enumeration of urban areas over 10,000 population and a quarterly sampling of the remainder of the country. SCS covered new dwelling unit completions only. It ignored alterations and extensions to existing buildings and dwelling units created by conversion. SCS covered only residential construction intended for year-round occupancy. In contrast, the census includes seasonal or marginal dwellings in the housing stock where residents have no other usual residence (although there were few of these as evidenced in Table 46). SCS also excluded collective dwellings. Finally, it included only certain types of dwellings, i.e. detached, semi-detached, row housing, flats, and apartments. It specifically excluded others, e.g mobile homes, trailers, tents, and boats. These other housing forms are included, however, in the census definition of the housing stock where residents have no other usual place of residence. The differing definitions employed in the census and in SCS add to the difficulty of constructing estimates of the components of housing stock change.

TABLE 47
Dwelling unit completions by type, Canada, 1951–80

	Type of dwelling				
Year	*Single detached*	*Duplex, semi-detached*	*Row*	*Apartment, other*	*Total completions*
1951–5	347,562	34,972	3,668	94,928	481,130
1951	60,366	7,568	585	12,791	81,310
1952	55,967	5,314	99	11,707	73,087
1953	68,916	7,714	372	19,837	96,839
1954	71,760	6,098	1,065	23,042	101,965
1955	90,553	8,278	1,547	27,551	127,929
1956–60	447,150	51,174	10,637	160,136	669,097
1956	95,656	11,872	2,137	26,035	135,700
1957	81,096	8,464	2,350	25,373	117,283
1958	96,830	10,004	2,226	37,626	146,686
1959	95,455	10,923	2,308	36,985	145,671
1960	78,113	9,911	1,616	34,117	123,757
1961–5	374,678	46,486	15,915	237,402	674,481
1961	76,171	10,593	2,019	26,825	115,608
1962	75,593	11,922	2,451	36,716	126,682
1963	71,585	7,150	3,487	45,969	128,191
1964	76,225	8,091	3,861	62,786	150,963
1965	75,104	8,730	4,097	65,106	153,037
1966–70	367,328	46,443	39,039	401,270	854,080
1966	73,858	7,707	6,412	74,215	162,192
1967	73,631	9,089	5,431	61,091	149,242
1968	74,640	10,098	7,896	78,359	170,993
1969	78,584	10,483	7,827	98,932	195,826
1970	66,615	9,066	11,473	88,673	175,827
1971–5	555,295	63,993	81,363	453,596	1,154,247
1971	82,978	12,518	16,795	88,941	201,232
1972	106,508	13,184	14,416	98,119	232,227
1973	122,696	13,479	14,832	95,574	246,581
1974	129,704	12,509	19,225	95,805	257,243
1975	113,409	12,303	16,095	75,157	216,964
1976–80	555,435	83,342	111,635	386,816	1,137,228
1976	128,623	15,160	21,172	71,294	236,249
1977	117,792	17,281	31,561	85,155	251,789
1978	106,195	19,155	26,644	94,539	246,533
1979	112,105	18,071	18,860	77,453	226,489
1980	90,720	13,675	13,398	58,375	176,168

SOURCE: *Canadian Housing Statistics*, 1976 and 1981

In Table 47 are presented data on housing stock completions between 1951 and 1980, as estimated from SCS. Annual total completions rose steadily in the first part of the postwar period. They crested in the 1970s and began to subside by 1980.

AGGREGATE GROWTH OF THE STOCK

Census data provide the best available measure of the total housing stock in Canada. However, censuses are intended to enumerate people, not dwellings. The usefulness of a census in counting the housing stock arises because the census uses the household as its unit of observation. For each household, there corresponds an occupied principal dwelling. In this sense, a census is effectively an enumeration of occupied principal dwellings. For our purposes, a major limitation of the census is that it does not extensively consider unoccupied dwellings, the latter including, in census terminology, seasonal and marginal homes. Nor does it extensively consider second homes.

Over the years, census takers have increasingly attempted to count unoccupied dwellings and, to a lesser extent, second homes. In published reports of the 1941 Census, no mention is made of unoccupied units. However, from 1951 onward (excluding the 1956 Census, which had no housing component), attempts were made to include some categories. The following definitions of a vacant dwelling were employed between 1951 and 1971. "This was defined as a dwelling suitable for occupancy, but unoccupied at the time of the Census" (*1951 Census of Canada*, vol 3, xviii). "Refers to any dwelling suitable for occupancy, but unoccupied at the time of the census. The term 'vacant' does not apply to dwellings temporarily closed or unoccupied summer cottages, cabins, shacks, trailers, etc" (*1961 Census of Canada*, vol 2(2), xviii); same definition given in the *1966 Census* (vol 2, xix-xx). "Refers to a dwelling, not a seasonal or vacation home, which was suitable and available for immediate occupancy, but which was not inhabited on Census Day. Newly constructed dwellings, completed and ready for occupancy, but as yet unoccupied at the census data were counted as vacant. This does not refer, however, to dwellings in which the occupants are temporarily away" (*1971 Census Dictionary*, 38). These definitions are broadly similar. In each, an important element is that the dwelling be available for immediate occupancy. Dwellings that were not immediately available for whatever reason were excluded from the count of total dwellings.

In 1976 and 1981, the notion of an unoccupied dwelling was expanded to include both available and unavailable units. The latter include unoccupied dwellings that are company maintained, a second home, rented or sold, to be demolished or expropriated, or unavailable for other reasons. As before, the 1976 and 1981 censuses did not include unoccupied seasonal or marginal dwellings; any second homes included had to be suitable for year-round occupancy.

This should serve to make the reader wary of any estimate of total housing stock. People can live in a wide variety of housing, including for example tents, trailers, and houseboats. Presumably a measure of the total housing stock could, or should, include all possible "dwellings" regardless of their form. However, it would be impractical to enumerate the entire housing stock within the terms of reference of a standard census, i.e. one focusing on an enumeration of persons rather than housing units. The 1976 and 1981 censuses broadened coverage of the housing stock but still used a narrowly defined criterion for identifying unoccupied "dwellings."

Mindful of this limitation, let us consider the estimates shown in the top row of Table 48.[2] Between 1951 and 1981, the total stock of dwellings in Canada went from 3.5 to 8.8 million units, an average annual increase of about 3.1 per cent. In contrast, Firestone (1951, 45) estimates that between 1921 and 1949 the housing stock grew at just 1.8 per cent annually.[3] Remember that some of the postwar increase was attributable to the more liberal definition of an unoccupied dwelling used in the 1981 Census. Also, after 1971, the census included dwellings occupied by temporary or foreign residents in Canada. Note also that collective dwellings formed a small, and relatively constant, portion of this stock. From published census reports, we know little about these categories of dwellings.

How close is the census increase in total dwellings to CMHC's data on new completions? The census increase in dwellings from 1 June 1951 to 30 May 1981 is 8.785 − 3.534 = 5.224 million units. From Table 47, total completions between 1 January 1951 and 31 December 1980 are 4.970 million units. Even allowing for the difference in time coverage, these figures are close. Given the differences noted above between the coverages of censuses and of the Starts and Completions Survey, the small difference between them may be accidental. Also, although the small difference suggests that the net effect of conversions, demolitions, and abandonments was small, it says nothing about the gross magnitudes of these components.

Because of some remarkable geographical shifts in Canada's postwar population, looking simply at aggregate figures can be misleading. The extensive rural-urban migration that marked the first half of the century was winding down in the 1950s. In addition, the 1950s and 1960s saw substantial net migration from smaller urban centres to metropolitan regions. The 1970s were somewhat different again, with emphasis on migration to western Canada and to smaller urban centres. However, much of the housing stock was immobile and could not be easily or economically moved to accommodate these population shifts. To a certain extent, population mobility thus meant abandoning housing in one part of the country and building replacement housing elsewhere. More new

TABLE 48
Dwellings (000s) by type and tenure, Canada, 1941–81

	*1941**	*1951*	*1961*	*1966*	*1971*	*1976*	*1981*
Total dwellings†	–‡	3,534	4,767	5,420	6,342	7,567	8,785
Private dwellings	–	3,522	4,745	5,402	6,325	7,551	8,757
Occupied by usual residents	2,576	3,409	4,555	5,180	6,035	7,166	8,282
By type of dwelling							
Single detached	1,853	2,276	2,979	3,234	3,592	3,992	4,735
Single attached	–	–	405	402	679	587	837
Double house	160	–	277	239	368	325	437
Row house	39	–	–	–	–	216	353
Attached to nonresidential	–	–	–	–	–	45	47
Flats and apartments	533	886	1,151	1,516	1,699	2,413	2,493
Duplex	–	–	403	340	407	351	400
Apartment	–	–	748	1,176	1,292	2,062	2,093
Movable homes	excl.	–	20	28	64	174	215
By tenure							
Owner-occupied	1,458	2,237	3,006	3,270	3,637	4,431	5,142
Rented	1,116	1,172	1,549	1,911	2,398	2,735	3,140
Occupied by temporary/foreign residents	excl.§	excl.	excl.	excl.	excl.	19	56
Unoccupied private dwellings#	–	113	190	222	290	366	419
Collective dwellings	–	12	22	18	18	16	28

SOURCE: Taken from published reports of the census of Canada, various years
* There are some comparability problems between different censuses. See text for details.
† Censuses in 1976 and 1981 include dwellings occupied by temporary or foreign residents.
‡ Data not available.
§ Category excluded in that census
# Censuses in 1951, 1961, and 1966 also included as vacant dwellings those that were still under construction. To maintain comparability with more recent censuses, such dwellings have been excluded from the figures presented above.

housing was needed than might have been expected simply by looking at aggregate net household formation.

It is difficult, however, to get an overall estimate of the impact of geographic mobility on housing abandonment and new construction. In part, this is because census definitions and boundaries have changed. What was defined to be a rural dwelling in 1951 may have been deemed "urban" in 1961. What had been a dwelling in a small town in one census may find itself within a metropolitan area in the next. We can find, from a census, how many dwellings were located in rural areas, or in urban areas of various sizes, but definitional and boundary changes severely limit comparisons between censuses.

One case study illustrates some aspects of how population decline was reflected in housing stock change. Teck Township in northern Ontario includes the town of Kirkland Lake and some outlying villages. During

TABLE 49
Dwellings by type, Teck Township, 1961 and 1966

Dwelling type	*1961*	*1966*
Single detached	2,094	2,246
Duplex flat	1,296	624
Other flat or apartment	1,129	1,549
Other dwelling	275	185

SOURCE: Taken from published reports of the 1961 and 1966 censuses of Canada

the late 1950s and the 1960s, the economic base of this township, gold mining, declined as the mines ran out of ore. Between 1961 and 1966, the population of Teck Township declined by 10 per cent. At the same time, the total number of occupied dwellings fell only about 4 per cent. Table 49 shows the distribution of occupied dwellings by type in Teck Township in the two censuses. Between 1961 and 1966, there was little new construction. The principal housing stock change occurred through conversion. The number of duplex units declined by one-half. Some of the duplexes were apparently converted to single detached; others were converted to triplexes or other types of apartment structures. Without good housing accounts data, one can only speculate on the extent of such conversions. However, there may well not have been any housing "abandonment" in the sense of entire structures being left vacant. Rather, the decline in population may well have forced a reduction in the number of basement or upstairs apartments as small landlords (occupying the main floors of buildings) withdrew such units from the market for lack of demand, expanding (perhaps involuntarily) their own living quarters.

The variety of ways that residential structures can be used adds to the difficulty of estimating the impact of geographic mobility on net housing stock change. Migrants do typically add to the demand for housing at their destination and hence to the number of occupied dwellings. However, their departure from the region of origin does not necessarily reduce the stock of dwellings there.

The next few sections examine changes in the characteristics of the housing stock. It should be noted at the outset, though, that published data are concerned almost exclusively with private dwellings occupied by "usual" residents. These dwelling units correspond exactly to the set of private households, the changing patterns of which are described in chapter four. Keeping in mind that we are describing just this subset (albeit a large subset) of the total housing stock, let us now examine some of its changing characteristics.

CHANGES IN TYPE OF DWELLING

Consider first the stocks of dwellings by type. Over the years, various censuses have employed somewhat different definitions of dwelling types. The data presented in Table 48 summarize comparable patterns of occupied dwellings by type.

In considering these, be mindful of enumeration errors in which a dwelling of one type is mistaken for another. Such errors can be commonplace. For example, what appears to be a single detached dwelling from the outside may contain a basement apartment. The structure should be classified as two duplex units but may be categorized as a single detached dwelling if the basement apartment is not detected. "There had been some misunderstanding on the part of the enumerators as to the application of the definition for single-attached dwellings. This misunderstanding appeared to be such that in 1941 too large a number of single-attached dwellings were reported. In 1951, the reverse appeared to be true, with enumerators tending to report too many apartments and flats" (*1951 Census of Canada*, vol 10, 362).

Changes in the way that censuses were conducted may also be significant here. Prior to 1971, enumerators conducted interviews. They were trained to be consistent in identifying types of dwellings, although, as noted above, errors still occurred. However, from 1971 on, censuses were based on self-enumeration. It was up to the respondent to identify the type of dwelling. In 1971, the respondent was given seven choices (eight in 1976), with a brief description of each. An accompanying instruction book provided drawings of typical dwelling units of each type. Problems were encountered with this self-enumeration approach, and a cautionary note was issued later with respect to 1971 counts. "Owing to a significant response bias in the 1971 figures for structural type of dwelling, particularly in the larger urban centres of Quebec, the comparability of 1976 and 1971 data should be viewed with caution. Different studies were undertaken to evaluate the extent of the error. It was found that in the identified problem areas in the Montreal core, the 1971 figures for 'single attached' were overstated at the expense of the 'apartment' category, which was underestimated by 36%" (*1976 Census of Canada*, vol 3.1, 31).

In the 1976 Census, a different approach was used. Before handing out each census questionnaire, the enumerator printed a dwelling type code on the outside of the form. The respondent still completed a question inside about dwelling type, and it was later compared with the enumerator's designation. Statistics Canada did not find it necessary subsequently to issue cautionary notes, and so this procedure appears to have been successful.

The 1981 Census, however, returned to the procedure used in 1971, the respondent being the sole source of information on dwelling type. Based on the 1971 experience, this might have been expected to create response biases. In addition, a decision was made to move the drawings of typical dwellings out of an accompanying instruction book and onto the questionnaire itself. The respondent was effectively faced with a variety of drawings and asked to pick the dwelling type most like his or hers. The danger was that someone living in an old house that had been flatted into a triplex, for example, might easily select the picture of a detached house rather than that of an apartment building. Apparently, response errors were commonplace. Statistics Canada enclosed the following warning in its some of its 1981 Census reports:

From the structural perspective the counts for Apartments in buildings with five or more storeys are believed to be relatively accurate. Counts for other types of dwellings in multiple unit structures (e.g. Apartments in buildings of less than five storeys and Row Houses), on the other hand, may contain varying degrees of error. For these dwellings there have been two types of misclassification. First, there are misclassifications among various types of the multiple unit structures. For example, Apartments in buildings of less than five storeys have frequently been classified as Row Houses, Semi-detached, etc. Second, there are some misclassifications between multiple and single structures. For example, a Duplex may have been misclassified as a Single Detached. "Cautionary Note on Data Quality - Structural Type" attached to *1981 Census of Canada: Occupied Private Dwellings: Type and Tenure*, Catalogue 92-903

With these evident limitations in mind, let us consider the structural type data presented in Table 48.

Single Detached Dwellings

In the 1981 *Census Dictionary* (p. 90), "Single Detached" is described as "a single dwelling not attached to any other building and surrounded on all sides by open space." Similar definitions were employed back to 1941. The definition excludes buildings where a portion (e.g a basement or an upper floor) has been flatted, such cases being duplex or apartment units. The single detached dwelling remained the principal form of housing in postwar Canada, although its importance declined. In 1941, single detached units formed 72 per cent of all occupied dwellings. By 1976, this had slipped to 56 per cent. In 1981, though, the incidence of single detached units began to climb again. Thus, in the postwar period, at least up to 1976, the traditional dominance of detached housing was challenged by other housing forms. Table 47 supplies corroborating evidence in terms of annual total housing completions.

It is of interest to compare census and completions data for single detached dwellings. Between 1951 and 1981, the census count of single detached dwellings rose by 4.735 - 2.276 = 2.459 million units. From Table 47, the number of completions in the calendar years from 1951 through 1980 was 2.647 million. Although similar in magnitude, an unknown number of new completions subsequently had basement and/or upstairs apartments added, converting them into duplex or apartment units in census terms. In addition, some detached housing was lost through fire, demolition, or abandonment. Again, there is little information on such changes. However, conversions into single detached dwellings likely played some role in expanding the supply. Some of this conversion was from duplexes, triplexes, or rooming-houses into single dwelling buildings. Another kind of conversion took the form of "winterizing" of seasonal homes, mostly used by retired people as an inexpensive or preferred form of housing.

Firestone (1951, 66) provides some contrasting historical information. He estimates that, during the 1920s, the proportion of dwelling units completed that were single detached fell sharply, from 72 per cent in 1921 to 47 per cent in 1929. During the Depression and into the early war years, single detached dwellings remained at about one-half of all completions. In the late years of the Second War, this proportion hovered at about three-fourths of all completions. The data in Table 47 indicate that single detached dwellings continued to form over 70 per cent of all completions through 1956. The period 1946-56 is thus unique; not since the early 1920s had single detached dwellings been such an important part of residential construction activity.

However, when the postwar period is viewed alone, the continued decline of the single detached unit up into the early 1970s is evident. What caused the decline? In part, one must take into account increasing urbanization within Canada. In 1941, only 55 per cent of the housing stock was urban, compared to 78 per cent by 1971.[4] In rural areas, single detached dwellings had long been the traditional housing form. In part, lower land costs there made detached housing more affordable. Also, rural areas had steady to declining populations, with little new housing construction; thus households tended to be housed in older, commonly single detached housing. It is in urban areas that one finds the widest variety of alternatives to single detached housing. Typically these housing forms trade off higher construction costs against reduced consumption of land. In larger urban areas, where land is relatively expensive, these types of residential development become more viable.

TABLE 50
Occupied dwellings showing percentage urban and percentage single detached by geographic area, Canada, 1941-81

Year	*Pct all occupied dwellings urban*	*Single detached dwellings as percentage of all dwellings*		
		All Canada	*Geographic area*	
			Rural	*Urban*
	(1)	*(2)*	*(3)*	*(4)*
1941	55.0	71.9	93.4	52.0
1951	63.2	66.8	90.2	53.1
1961	72.0	65.4	90.0	55.9
1966	76.1	62.4	91.1	53.4
1971	78.5	59.5	89.2	51.4
1976	78.3	55.7	84.2	47.8
1981	78.6	57.2	85.1	49.7

SOURCE: Computed from published census reports, various years

Table 50 gives an indication of the role of increasing urbanization. The left hand column, (1), shows the percentage of dwellings that were in urban areas. The next column, (2), shows the percentage of all dwellings in Canada that were single detached. Columns (3) and (4) give the percentage of urban and rural dwellings, respectively, that were single detached. Note that column (2) is the weighted average of columns (3) and (4), the weights being determined by column (1).

Between 1941 and 1966, single detached dwellings declined as a percentage of the total stock; see column (2). However, in rural and urban areas taken separately, there were only small changes. Almost all of the relative decline of single detached dwellings was tied to the rural-urban transition. Increasing urbanization, with its distinctive housing forms, played a major role in the declining importance of single detached housing between 1941 and 1966.[5]

However, an "increasing urbanization" hypothesis does not explain the decline of single detached housing between 1966 and 1976. The level of urbanization, as indicated in column (1), increased only modestly. Looking at columns (3) and (4), we see instead the decline of single detached housing in both rural and urban areas. Something other than a rural-urban shift was involved. What?

One possible explanation relates to the impact of the maturing postwar baby boom. Beginning in the mid- to late 1960s, the number of young adults seeking separate accommodation began to boom. Few of these people had the wealth necessary to make a large downpayment on an owner-occupied dwelling. Traditionally available primarily to owner-occupiers, the single detached dwelling may have been the housing form preferred by many but affordable to few. Rental housing, or less expensive forms of owner-occupied housing, dominated during this period. Such a demographic argument can also be used to explain the upturn in single detached housing between 1976 and 1981. As the baby boomers aged, their incomes rose, and they accumulated money for a downpayment. This put more of them in a position to purchase single detached dwellings than had been the case. Such an argument suggests that the 1976–81 period might mark just the beginning of a period, of a decade or two, during which single detached units become an increasingly large part of the occupied housing stock.

Another demographic argument focuses on the changing patterns of family formation in Canada. L.B. Smith (1971, 12), among others, has noted that family households are much more likely to occupy single detached housing than are other types of households. The rapid rise of nonfamily households thus helps to explain why single detached housing became less prevalent between 1966 and 1976. In 1966, only 16 per cent of all households in Canada did not contain a family; 10 years later, this had risen to 21 per cent.

At the same time, however, the propensity for a family, or nonfamily, household to live in a single detached dwelling fell. Only 31 per cent of nonfamily households lived in single detached dwellings, compared with 42 per cent a decade before. Among family households, the decline was smaller, from 66 per cent to 62 per cent. Thus, although the emergence of nonfamily households was important in shifting the dwelling type distribution, there would have been a shift even if the relative numbers of family and nonfamily households had not changed.

The above arguments suggest that demographic changes and a prosperous economic environment encouraged the changing mix of dwelling types observed between 1966 and 1981. However attractive this explanation, one should not lose sight of other important considerations. One has been the evolution of housing design. In part, this evolution was motivated by the rapid escalation of house prices in the early 1970s, which made the promise of affordable home ownership appear less achievable than ever. A variety of new forms of single detached housing emerged. Let us consider just a couple of illustrations. One was the "zero lot line," which eliminated some of the traditional minimum sideyard and backyard setbacks; this gave the builder more flexibility in placing buildings on lots and effectively reduced typical lot sizes and building costs. Another was the "linked housing" concept. In linked housing, two or more dwellings are constructed on the same basement foundation. Above ground level, they are single detached units. Below ground level, they each have a separate basement with one or more party walls. The savings associated with linked housing are mainly a result of the reduced cost of construction of common foundations and the typically smaller lot sizes possible.

There were substantial changes during the postwar period in the styles of new single detached dwellings. CMHC (1971, 31-2) describes the changes between 1946 and 1970 as follows: "In the first few years after the war . . . the storey-and-a-half house . . . had a great revival . . . Ten years later, when home-buyers had more money in the bank to buy a bigger lot, the one-floor 'ranch' bungalow was preferred . . . From this, it was only a step to that ingenious invention of the 'fifties, the 'split level', which brought the basement space into the circulation of the house." During the 1960s, the popularity of the bungalow and split-level began to wane. Particularly in larger urban areas, where land was more expensive, there was a return to the two-storey dwelling.

Single Attached Dwellings

"Single attached" includes all dwellings that are in single-dwelling buildings attached on one or more sides to other dwellings or nonresidential buildings. It includes double houses (also known as semi-detached units), row houses (three or more dwellings in a row), and dwellings attached to nonresidential structures. It specifically excludes vertically stacked dwellings, these being treated as apartments. It excludes also dwellings situated over business or other nonresidential premises, these also being treated as apartments. From 1961 to 1971, the number of single attached dwellings rose sharply: over 5 per cent annually. Between 1971 and 1981, however, their numbers increased only modestly.

Note also, though, the variability of these counts. In both 1966 and 1976, the counts of single attached dwellings were each lower than in the preceding census. What happened? Were there widespread demolitions, abandonments, or conversions of attached units between 1961 and 1966 (and between 1971 and 1976)? It is difficult to answer such questions without good housing accounts data. We simply do not know what role these factors played, if any, in the stock reductions observed during this period.

We have already discussed the enumeration problems with regard to dwelling type in the 1971 and 1981 censuses. To an extent, the 1966 and 1976 counts could appear to be "off" because of enumeration errors. It is difficult to assess the significance of this, however, from published census reports.

Published data on completions do not identify the number of completions of attached dwellings. In Table 47, attached units would include the column marked "Row" housing and part of the column entitled "Duplex, semi-detached." Completions of row housing totalled 0.248 million units in the period 1961-80, completions of duplex and semi-detached housing an additional 0.240 million. Completions of single attached dwellings (which would also include a small number of dwellings attached to nonresidential structures) therefore must have totalled less than 0.5 million units. The number of single attached units enumerated in the census rose by 0.837 - 0.405 = 0.432 million units. To the extent that they are comparable, the completions data are roughly consistent with the change in census counts.

The completions data in Table 47 indicate how attached housing came to be an important component of Canada's housing stock. First, consider row housing. After a low level of completions in the early 1950s, the annual rate settled around 2,000 units through 1962. During the period 1963-77, completions of row housing climbed quickly, eventually reaching 30,000 units annually. From 1978 to 1980, however, the decline in row completions was abrupt. Thus it is the period from the mid-1960s to the mid-1970s that saw the greatest rate of construction of row housing. The peak in duplex (which, to repeat, are not "attached" units as defined here) and semi-detached (which are) unit completions occurred during the 1970s, reaching an average of about 14,000 per year. Thus, the boom in duplex and semi-detached units overlapped and then superseded the boom in row housing.

What caused these booms? One explanation is that they reflect increasing urbanization. As noted above, these housing forms are distinctly urban and their numbers should mirror the growth of cities relative to rural areas. However, this argument does not accord well with the facts. As noted above, postwar urbanization was largely completed by the mid 1960s, and yet the peak in construction of attached dwellings appears to have occurred later. Rather, I think that these housing forms grew in response to the economic and demographic pressures of the time. The maturing of the baby boomers created strong demands for affordable housing. Typically, semi-detached and row housing was less expensive to construct and to heat. With the emergence of condominium ownership in the late 1960s and the early 1970s, it became more feasible, as well as economical, to construct row housing.

Flats and Apartments

"Flats and Apartments" include all dwellings that are in duplexes, triplexes, double duplexes, or apartment buildings containing three or more units. It includes also dwellings situated over business or other nonresidential premises. Remember that duplexes include any detached building that contains a flatted basement or upstairs. A detached building containing three flats (triplex) or more is treated as an apartment

TABLE 51
Occupied dwellings showing percentage apartments or flats by geographic area, Canada, 1941–76

		Geographic area	
Year	*All Canada*	*Rural*	*Urban*
1941	20.8	3.0	35.4
1951	26.0	4.7	38.4
1961	25.3	3.3	33.8
1966	29.3	3.4	37.4
1976	33.7	4.7	41.7

SOURCE: Computed from published census reports, various years

house. The census count of occupied apartments and flats rose steadily from 1951 to 1976. Separate data for duplex units, available only from 1961, show that the number of such units stayed constant. The number of apartments, however, rose steadily. By 1981, the growth of apartments and flats appeared to have ended abruptly.

Published data on completions do not allow us to identify the total completions of apartments and flats. In Table 47, these would include the column marked "Apartments, other" and part of the column entitled "Duplex, semi-detached." Nevertheless, as a proportion of total completions, apartments were apparently most important during the late 1960s and the early 1970s. Between 1969 and 1974, apartment completions averaged 90,000 annually, double the level experienced just six years before and higher than anything experienced since (except in 1978).

What lay behind these changes in apartment construction? One possible answer, as before, is that it reflects increasing urbanization. As with row and semi-detached housing, apartments tend to be an urban phenomenon. Table 51 shows the percentage share of apartments or flats in rural areas, urban areas, and all Canada between 1941 and 1976. Data for 1971 and 1981 are omitted because of enumeration problems. Between 1941 and 1966, the period during which most of the postwar urbanization occurred, apartments and flats nationally rose from 21 per cent to 29 per cent of all occupied dwellings. At the same time, there was no marked trend in this percentage within either rural or urban areas, suggesting that increased urbanization brought about the change at the national level.

However, the national change between 1966 and 1976 appears to be attributable to the fact that an increasingly higher proportion of urban dwellings were apartments or flats. A common view is that the apartment boom was another response to the economic and demographic pressures created by the aging baby boomers. For many young adults, an apartment or flat was seen as a "starter" housing unit. These units were typically rented and involved lower (out-of-pocket) costs for persons unable to make a substantial downpayment on owner-occupied housing. There is evidence that the aging baby boom helped to create the apartment boom. The timing of the boom, in particular the way that it was followed by the row housing boom and then the single detached housing boom, is indirect evidence of upmarket movement, as households moved from apartments to smaller (often attached) housing, to (typically larger) detached housing.

Movable Homes

Postwar censuses recognize two kinds of movable dwellings. One was a dwelling such as a tent, travel trailer, railroad car, or houseboat. The other was the mobile home, i.e. a dwelling designed and constructed to be transported on its own chassis and capable of being moved on short notice. Mobile homes make up the vast majority of movable homes, so much so that the terms can virtually be used interchangeably. Note, though, that this category excludes many manufactured homes. Manufactured dwellings that are transported to a given site on a trailer and then set up on a foundation are treated as single detached dwellings, not movable dwellings.

Although never common, movable dwellings were the fastest-growing form of housing in postwar Canada. Table 52 shows movable dwellings as a percentage of all occupied dwellings by geographic locale. Mobile homes were predominantly found in rural nonfarm areas, forming 8 per cent of the occupied stock there in 1976. The relative absence of mobile homes in larger urban areas is evident. At the same time, note the rapid increase in all geographic locales, between 1961 and 1976, in the incidence of movable dwellings.

Why did the incidence of movable homes rise during this time? Several factors come to mind. One is that such dwellings became less expensive, relative to other forms of owner-occupied housing. Another is that such dwellings were well suited to the needs of a more mobile population. Particularly for persons involved in activities such as resource development and highway construction, the ability to be able to set up a home in a new location and to move it when one's job was relocated must have been an attractive feature. A third explanation might concern technological changes in the construction of mobile homes, which have made such dwellings better suited to the harsh Canadian climate.

TABLE 52
Occupied dwellings showing percentage of movable dwellings by geographic area, Canada, 1961-76

Geographic locale	*1961*	*1966*	*1971*	*1976*
Rural areas				
Farm	0.2	0.4	1.0	3.5
Nonfarm	1.4	1.8	3.6	8.3
Urban areas				
Under 5,000 population	0.8	1.1	2.1	4.2
5,000 persons or more	0.2	0.2	0.4	0.8

SOURCE: Computed from published census reports, various years

CHANGES IN TENURE

Every census since 1941 (with the exception of 1956) has included a question on dwelling tenure. Only two tenure states were admitted, owning and renting, although the 1981 Census enumerated condominium ownership separately. Counts of occupied dwellings by tenure are presented above in Table 48. From these data can be seen the rise in ownership between 1941 and 1951, a stable propensity to own during the 1950s, a decline in the 1960s, and a slow rise in ownership during the 1970s.

It is instructive to contrast the postwar period with earlier times. Firestone (1951, 51-2) reports that about two-thirds of all dwellings were owner-occupied in 1921. This is slightly above the 62 per cent figure for 1981. Further, Firestone estimates (274) that owner-occupied dwellings declined as a percentage of all dwellings (to about 56 per cent by 1943) and then began to increase. Thus the early postwar rise in owner-occupancy seen in Table 48 really began in the late war years. Further, Firestone's data make the early postwar period and the 1970s, with their rising levels of owner-occupancy, seem anomalous. Why was there a general decline in owner-occupancy in the prewar years? What altered this trend in the early postwar period and again in the 1970s?

Before looking at these temporal changes, it helps to consider the definitions of the two tenure states. Postwar housing markets can be characterized by their increasing complexity. We have already commented on some of the new structural forms that emerged. New legal forms emerged as well, with implications for the definition of tenure.

Except in 1951, each census since 1941 has defined tenure in the same broad terms. If the dwelling is owned by the head or any other member or members, the household is said to be an "owner occupier."[6] This includes cases where there is a mortgage or other lien outstanding on the property. All other households are treated as "renters." Note that some renters paid no cash rent or a reduced rent, e.g. persons living free in accommodation owned by relatives or friends, or employees who are paid in part or whole with free housing services.

Later censuses attempted to clarify further this definition: "A dwelling is classified as owned even though it may be . . . in the process of being bought (e.g. closure has not yet been finalized) . . . A dwelling which is rented with an option to buy is considered to be rented until the option is taken up" (*1976 Census of Canada: Public Use Sample Tapes: User Documentation*, 189). "A dwelling [which] is classified as 'owned' . . . may be situated on rented or leased land or be part of a condominium (whether registered or unregistered). A dwelling is classified as 'rented' . . . if the dwelling is part of a co-operative. For census purposes, in a co-operative all members jointly own the co-operative and occupy their dwelling units under a lease agreement" (*1981 Census Dictionary*, 78).

These clarifications are a response to the emergence of condominiums, co-operatives, and lot leasing (particularly in conjunction with mobile homes). While perhaps initially clear, the distinction between owning and renting can become less evident. An example is the 99-year "net-net" lease, wherein the lessee pays an upfront fixed sum for the lease and an annual amount for property taxes, utilities, insurance, and so on. If the lessee is free to sublet at any time, the upfront sum paid to the original lessor is not unlike the purchase price of an owner-occupied dwelling, in magnitude and purpose. In effect, the lessor gives up the right to increased rental value over the life of the lease in return for a guaranteed sum (the upfront payment). The lessee gains this important aspect of the owner's property rights. Nevertheless, the census tenure definitions would treat the lessee as a renter.

The 1940s

Remember that 1941 was in the middle of the war. With many scarce resources being allocated to the war effort, there was little material or manpower with which to build new housing. Also, wartime rent controls dampened the links among demand, price, and new construction. Further, much of the existing housing stock was in need of major repair after the enforced neglect of the Depression. It is not surprising, therefore, that the prosperity of the immediate postwar period translated into substantial renovation and new home building. What is surprising, though, is just how fast the tenure distribution changed – from 57 per cent owners in 1941 to 65 per cent in 1951 – at a time of rapid urbanization. The move to cities should have brought about a decline in home ownership, other things being equal. The decline in single detached dwellings, as noted in the previous section, is certainly consistent with this argument. Why, then, did the incidence of ownership not decline?

A clue can be found in the 1951 Census, where the following is noted: "Accompanying an apparent increase in owner-occupied single homes, there has been a definite swing away from tenant-occupied single dwellings in the last ten years. Indeed it is estimated that there has been a drop of at least 10 per cent between 1941 and 1951 in the number of tenant households living in single dwellings in urban areas" (*1951 Census of Canada*, vol 10, *General Review*, 360). This drop implies a substantial conversion of rented detached dwellings during, or shortly after, the war. Conversions to what? Again, without proper housing accounts data, one cannot be conclusive. However, many of these conversions likely were to owner-occupancy. In other words, some of the single detached dwellings that had been rented in 1941 were sold to owner-occupiers by 1951.

The scale of this tenure conversion should not be surprising. During the Depression, house prices tumbled. Foreclosures were commonplace as home owners found themselves unable to keep up fixed mortgage payments.[7] This deflation must have been a disincentive to sell for many existing owners, as well as for the banks and others that obtained ownership upon foreclosure. Housing market conditions did not improve much until the later war years. Existing owner-occupiers who had to relocate or, for other reasons, vacate their own dwelling were not inclined to sell their property under such conditions. To the extent that they believed that house prices would rise in the future, it would be rational for them to hold on to that dwelling, renting it out over the short term. For this reason, the incidence of ownership in 1941 was abnormally low. In the buoyant period immediately following the war, these landlords found willing buyers for their properties, and the buyers were largely owner-occupants. The release of this "pent up" supply of tenant-occupied single detached housing into the owner-occupier market accounts for an important part of the tenure conversion evidenced by 1951.

There are two important points in the above argument. First, certain kinds of tenure conversion are relatively costless. In the preceding section, we discussed conversions between dwelling types, e.g. single detached to duplex structures. Conversions between dwelling types usually involve at least some carpentry work, and hence some expense. However, to convert a single detached dwelling from owner- to renter-occupancy may well involve an even smaller (or perhaps no) expense. Some other kinds of tenure conversion, though, are more difficult. For example, it is not a simple matter to convert rental apartments to owner-occupancy. In the 1970s, this was sometimes accomplished by converting a rental apartment building into a condominium. However, condominium conversions have generally required government approval, and

TABLE 53
Owner-occupiers as a percentage of all households by geographic area, Canada, 1951-61

		Geographic area	
Year	*All Canada*	*Rural*	*Urban*
1951	65	82	56
1961	66	83	59

SOURCE: Computed from published census reports, various years

there was some opposition to such conversions where rental housing was thought to be in short supply.

Second, the above speculation emphasizes the importance of specific economic considerations in shaping tenure. Broad demographic changes, including increasing urbanization, may be important determinants of tenure change. However, they can not be considered in isolation from the specific economic conditions of the times.

The 1950s

In contrast to the preceding decade, the period 1951-61 saw almost no change in tenure patterns. In 1961, 66 per cent of households were owners, compared to 65 per cent in 1951. If the change between 1941 and 1951 was an adjustment from the Depression and war years to a "more normal" prosperous period, the adjustment may have run its course by 1951, and the lack of subsequent change need not be surprising. However, the 1950s were a period of rapid urbanization. Did this urbanization have no effect on tenure patterns? The data in Table 53 help us to understand what happened. There was a shift from rural areas (where over 80 per cent of households are owners) to urban areas (where under 60 per cent are owners). However, this was compensated for by the rise in the incidence of ownership, especially in urban areas.

What caused the substantial rise in the incidence of urban home ownership? We return to this question in chapter eight where it is argued that the increasing prosperity of households during the 1950s made ownership more widespread than before.

The 1960s

The 1960s contrast strongly with the two earlier decades. Whereas earlier the incidence of ownership had been increasing, it now began to fall, from 66 per cent owners in 1961 to 63 per cent in 1966 to 60 per cent in

TABLE 54
Owner-occupiers as a percentage of all households by type of dwellings, Canada, 1961-71

		Type of dwelling		
Year	*All Canada*	*Single detached*	*Single attached*	*Flat/ apartment*
1961	66	86	50	19
1966	63	87	49	16
1971	60	86	38	13

SOURCE: Computed from published census reports, various years

1971. What brought about this change? In the 1950s, the incidence of ownership rose only slightly because the negative effect of increasing urbanization was just offset by increasing affluence. During the 1960s, these two trends continued. Especially marked was the spatial shift of population into Canada's three major metropolitan areas.

Historically, large cities in Canada had much greater incidences of renting. In part, this is because of higher land prices. Households that may have been able to afford ownership in a smaller town found it unaffordable in metropolitan areas. Instead, they turned to types of housing - e.g. apartments and townhouses - that made more intensive use of land and hence were less expensive. For the most part, such housing could only be rented.

Table 54 illustrates how the tenure shift occurred. It shows, for 1961, 1966, and 1971, the percentage of dwellings that were owner-occupied, by dwelling type. Over this period, there was a shift toward apartments, semi-detached, and row housing. Single detached housing fell from 65 per cent of all occupied housing in 1961 to 59 per cent in 1971. The shift away from single detached dwellings itself helps to account for the decline in ownership, because the other housing forms were predominantly rental units. Of course, this is not an entirely satisfactory answer. There need not necessarily be a connection between dwelling type and tenure. Apartments and townhouses, for example, can be owned as well as rented. However, typically this is done via condominium ownership. Until the late 1960s, few provinces had legislation enabling condominiums.

However, Table 54 indicates that the decline in overall ownership was due also to a declining incidence of ownership of apartments or flats between 1961 and 1971 and a decline in ownership of single attached units between 1966 and 1971.[8] What provoked these? Let us consider first apartments and flats. Traditionally, an important source of rental housing was the duplex or triplex structure, where the landlord lived in one flat and rented out the remainder. Even in 1961, for example, just over one-half of all apartments and flats were in buildings containing just two or three dwellings. This meant that as many as one out of perhaps two or three apartments or flats was owner-occupied. However, the postwar period marked the era of the large apartment house with many tenants and, typically, a nonresident or corporate landlord. It was the relative decline of the duplex and triplex with their resident landlords that accounted for the decline in apartment/flat ownership between 1961 and 1971. This in turn may also be associated with the shift out of smaller towns (and smaller buildings) into metropolitan areas (and larger apartment buildings).

In the case of single attached housing, a different argument is needed. Traditionally, such housing was primarily owner-occupied. Two forms, the semi-detached and row units, were predominantly found in the older areas of larger cities. Built originally in response to high land costs, they had widely been owner-occupied from the outset. Another single attached form, the house adjoining a store, had often been the home of the small proprietor, with a corresponding high incidence of owner-occupancy. In the 1960s, however, the nature of this housing form began to change with the construction of suburban townhousing. Some groups of townhouses had street frontage; i.e. each townhouse had its own access to a public street. In such cases, ownership was possible. However, many were constructed in "estates" or clusters on private land, facing only private roads. Until the emergence of condominium ownership in the 1970s, many of these "estate" townhouses had to be rental units, because of the lack of direct access to a public thoroughfare.

Other factors also contributed to the decline in owner-occupancy in the 1960s. One was the large typical family size, a product of the postwar baby boom. As a consequence, modest-income families found it difficult to acquire the savings for a downpayment that would make ownership affordable. Another was the maturing of the leading edge of the baby boom. In the last half of the 1960s, the baby boomers were reaching early adulthood and beginning to form their own households. In spite of widespread increasing affluence, this cohort simply could not generally afford ownership, however desirable it was to them. They typically needed time to work and save for an adequate downpayment. Also, the 1960s marked the beginning of the marriage bust and the rise of the nonfamily individual. Nonfamily households traditionally had been predominantly renters, and this accounts for some of the decline in ownership in the 1960s.

Chapter nine considers the role of the public sector in housing provision. It is argued there that assisted housing, in its many different forms, essentially began on a large scale in the 1960s. Much of this assisted housing was rental and was geared to individuals (particularly the elderly) and families who might not otherwise have been able to afford separate housing in the private sector. The result was a rapid rate of household formation, concentrated among low-income groups and almost entirely rental. This rise in assisted housing also accounts for some of the decline in ownership during the 1960s.

The 1970s

During the 1970s, the incidence of ownership began to climb again, from 60 per cent in 1971 to 62 per cent in 1981. Ownership levels were still below the peaks of 1961, but the drop of the 1960s had been reversed. In certain respects, the first and last half of the 1970s are different. The data in Table 55 give the incidence of ownership by dwelling type for 1971, 1976, and 1981. Between 1971 and 1976, single detached dwellings declined as a proportion of the occupied housing stock; later they rose. Other things being equal, those dwelling type shifts should have produced a decline in ownership in the first half of the 1970s and a rise in the last half. However, the incidence of ownership rose even in the first half. Why? Apparently, there was a substantial increase in the incidence of ownership among dwellings that were not single detached. In marked contrast to the decline in ownership in such dwellings in the 1950s and 1960s, this likely reflects the condominium boom of the early 1970s. However, it is not possible to verify this from published census reports, as these do not distinguish between condominium and other kinds of ownership. Note also that the incidence of ownership of "other" dwellings did not increase in the last half of the decade. In part, this reflected the subsequent "bust" in the condominium market.

The rise of owner-occupancy in the 1970s was fuelled at first by the emergence of condominium ownership and later by the shift back to single detached housing. Why did these changes happen? One view is that they reflected the changing housing preferences, and ability to afford, of the aging baby boomers. In the early 1970s, the condominium unit typically offered low-cost entry to the owner-occupied market. To many young households, this must have been an attractive option in view of the then-rapid escalation of house prices. Such a purchase was often seen as a toehold in the real property market. Perhaps enticed from a rental apartment to a low-cost condominium unit, the typical baby boomer continued to save and plan for a larger house. Single detached housing continued to be the preferred housing form for most family households, and many began to make the switch in the mid- to late 1970s.

These are not, however, the only considerations. One should also remember the role of government in promoting condominium ownership in the early 1970s. Financial assistance was provided to moderate-income

TABLE 55
Owner occupiers as a percentage of all households by type of dwelling, Canada, 1971-81

Year	Type of dwelling	
	Single detached	*Other dwelling*
1971	86	22
1976	90	26
1981	89	26

SOURCE: Computed from published census reports, various years

households to enable them to purchase a first home. In certain cases, the assistance was substantial enough, and the financing generous enough, to make the out-of-pocket costs of ownership lower than the cost of renting. Some of the increase in condominium ownership was attributable to such assistance. See chapter nine for more details.

To understand the rise of single detached housing in the late 1970s, it is important to keep in mind the role of the condominium "bust." Condominium housing, having been newly introduced in most provinces, underwent growing pains. After the initial euphoria about condominiums as a new life-style that also promised property value appreciation came the inevitable letdown. Shoddy construction, poor management, and inadequate financing in certain cases created problems for condominiums owners. With some exceptions in the luxury condominium stock, many owners found that their properties were not appreciating in value or, worse, had fallen and that potential purchasers were shying away because of the publicity given to these problems. These problems served not only to discourage extensive new condominium development in the late 1970s; they also encouraged some existing condominium owners to get out as soon as possible by speeding up plans to purchase other forms of owner-occupied housing.

CHANGES IN THE SIZE OF DWELLING

As discussed in chapter two, the size of a dwelling may be measured in different ways, e.g. number of rooms or floor area. However, census data are quite limited; there is no information, for example, on floor area. The only widely available measure is total number of rooms, and at that only decennially. Further, as described in chapter two the census definition of a room has varied over the years.

TABLE 56
Percentage distribution of occupied dwellings by number of rooms, Canada, 1941-81

Number of rooms	*1941**	*1951*	*1961*	*1971*	*1981*
One or two	11.0	6.8	4.6	4.7	4.0
One	4.0	1.9	1.4	1.5	-
Two	7.0	4.9	3.2	3.2	-
Three or four	29.4	29.7	28.5	28.1	25.2
Three	11.7	10.1	9.2	10.1	-
Four	17.7	19.6	19.3	18.0	-
Five or six	38.1	39.1	45.1	41.7	39.1
Five	18.0	19.7	23.9	22.6	-
Six	20.1	19.4	21.2	19.1	-
Seven to nine	18.8	21.1	19.1	22.3	27.1
Seven	10.3	10.9	10.2	12.0	-
Eight	6.1	7.0	6.2	7.1	-
Nine	2.4	3.2	2.7	3.2	-
Ten or more	2.7	3.3	2.7	3.2	4.6
All occupied dwellings	100.0	100.0	100.0	100.0	100.0
Average dwelling size†	5.3	5.3	5.3	5.4	5.7

SOURCE: Computed from published reports of the Census of Canada, various years
* In the 1941 Census, data are for households by number of rooms occupied. Unlike in later censuses, a household is a housekeeping unit and there could be more than one household per dwelling. In such cases, each room in the dwelling was allocated to a household. Therefore the number of rooms occupied by a household could be smaller than the number of rooms in the occupied dwelling. In later censuses, these two measures were identical by definition
† Average number of rooms in an occupied dwelling. In 1941, the average number of rooms occupied by a household. For 1941, the average number of rooms per occupied dwelling was 5.5.

It is commonly argued that a household's choice of dwelling size is limited by its income. We each might want a large, or larger, dwelling but find that it is simply too expensive given our preferences and income. Given that the postwar period was marked by generally increasing prosperity, was there a corresponding increase in the typical size of dwelling occupied? Interestingly, over the postwar period, average dwelling size did not change much. In the 1941 Census, the average number of rooms occupied per household was 5.3. It stayed about the same through 1971, and then rose to 5.7 rooms in 1981.

However, focusing on averages hides important changes in size distribution. In Table 56 are presented data on dwellings by number of rooms since 1951 and on households by number of rooms occupied in 1941. Between 1941 and 1961, the relative number of small dwellings (i.e. three or fewer rooms) declined. The relative number of four- and five-room dwellings rose sharply, and the number of larger dwellings remained roughly stable. Thus, although the average size of dwelling did not change, there was a bunching of the size distribution about the average.

From 1961 to 1981, though, the average size of dwelling began to increase, at first slowly and then more rapidly. As seen in Table 56, most of this increase came about not because of continued decline in small dwellings but from the emergence of dwellings of seven rooms or more.

Surprisingly, these changes in size distribution are not linked in a simple way to observed changes in dwelling type. In conventional wisdom, detached dwellings are typically thought to be larger than other kinds of dwellings. The average size of dwelling climbed in the 1970s at the same time as the incidence of single detached dwellings began to increase. However, the incidence of single detached dwellings fell steadily from 1941 to 1971, while the average size of dwelling stayed the same or even increased. What happened?

In part, the size changes were related to tenure shifts. Above, it is noted that the split between owners and renters remained roughly constant between 1951 and 1961, with the balance tipping to renters in the 1960s. Data from the 1941, 1961, and 1971 censuses suggest that the average size of a rented dwelling remained constant at about 4.4 to 4.5 rooms. However, the average size of an owner-occupied dwelling rose from 5.8 rooms in 1961 to 6.1 ten years later. This increasing average size of an owner-occupied dwelling was more than enough to offset the impact on average dwelling size of a shift toward rental accommodation.

This indicates a widening gap during the 1960s between renters and owners. In part, the growing divergence had to do with the aging baby boom. Crudely put, the divergence was partly between the baby boomers and their parents. The parents' generation translated the prosperity of the 1950s and 1960s into demands for larger, typically owner-occupied housing. Their adult children, having newly formed households and first starting jobs and forming families were not able to afford more spacious accommodation and instead tended to smaller, rented units.

To understand the growing tenure differential in dwelling size in the 1960s, one must also look to the role of assisted housing, particularly senior citizen housing. For the most part, this consisted of relatively small units. In the case of senior citizen housing, it consisted mainly of bachelor and one-bedroom units. Since much of this housing was rental, it undoubtedly played a role in keeping the average size of a rental unit from increasing similar to owner-occupied units.

The above arguments relate changes in dwelling size by tenure indirectly to changes in household size. Small households are seen to choose rental accommodation, e.g. young singles and couples and elderly widows and couples. However, the data in Table 57 suggest that dwelling sizes increased relative to the sizes of households that occupied them. This table shows, for each size of household, the dwelling size distributions

TABLE 57
Percentage distribution of occupied dwellings by number of rooms by size of household, Canada, 1941–81

Number of rooms	*1941**	*1951*	*1961*	*1971*	*1981*
(a) One-person households					
One or two	48	32	25	23	16
Three	14	18	20	27	29
Four	12	17	18	19	23
Five	8	12	14	12	14
Six or more	17	22	24	18	18
All dwellings	100	100	100	100	100
(b) Two-person households					
One or two	16	9	5	4	2
Three	19	16	17	17	10
Four	21	24	25	26	23
Five	16	20	22	22	24
Six or more	28	32	31	31	42
All dwellings	100	100	100	100	100
(c) Three-person households					
One to three	23	16	13	9	4
Four	21	24	25	24	15
Five	20	22	25	26	25
Six	19	19	20	20	22
Seven	8	9	8	11	15
Eight or more	8	10	9	11	19
All dwellings	100	100	100	100	100
(d) Four- or five-person households					
One to three	14	10	6	3	1
Four	18	20	18	13	6
Five	20	22	29	27	21
Six	25	23	25	24	23
Seven	12	12	11	16	19
Eight or more	12	13	11	17	30
All dwellings	100	100	100	100	100
(e) Six- or seven-person households					
One to four	23	21	14	8	4
Five or six	45	42	52	47	34
Seven	15	16	15	19	20
Eight or more	17	21	19	27	42
All dwellings	100	100	100	100	100
(f) Eight persons or more					
One to four	18	16	12	7	6
Five or six	39	34	39	38	31
Seven	16	16	19	21	18
Eight or more	27	33	30	33	46
All dwellings	100	100	100	100	100

SOURCE: Computed from published reports of the Census of Canada, various years
* See asterisked note to Table 56.

from 1941 to 1981. As an example, consider persons living alone, i.e. panel (a). In 1941, about one half of such persons lived in dwellings of just one or two rooms; just 17 per cent were in dwellings of six rooms or more. Moving across the columns from left to right, one sees how typical dwelling size changed from 1941 to 1981.

The following patterns are evidenced in this table. One is an abrupt decline in the incidence of small living quarters. What is "small" is relative to household size. Among persons living alone, "small" is a one- or two-room dwelling; among three-person households, decline is evidenced for up to five-room units. Among persons living alone, a second pattern is the convergence over time toward a typical dwelling size of from three to five rooms. By 1981, such dwellings accounted for about two-thirds of all persons living alone. After rising in the early postwar years, the proportion living in dwellings of six rooms or more fell after 1961. The relative decline of the large dwelling may have had a demographic basis. As noted in chapter four, among those living alone have traditionally been a substantial number of elderly or widowed, many of whom live in the larger dwellings in which they had raised their families. However, the influx of baby boomers meant that an increasing proportion of persons living alone were being housed in apartments or other dwellings with fewer than six rooms. A third pattern, among larger households, was the shift toward larger dwellings over time. The change was especially marked in the 1970s but shows up in earlier decades as well.

A common measure of dwelling adequacy is the number of persons per room. This is often taken to be an indicator of "crowding" in the dwelling. However, such a measure is quite sensitive to the mix of households, since smaller households tend to have fewer persons per room. Also, the problems of changing room definition and size variations make exact comparisons difficult. Mindful of these pitfalls, we can see from Table 57 that, overall, the number of persons per room fell sharply over the postwar period. Households of virtually all sizes were increasingly inhabiting larger and larger dwellings.

QUALITY CHANGE

To what extent did the quality of the housing stock improve (or worsen) in the postwar period? This is a fascinating question, because it is really asking about the improvement in an important aspect of social well-being. We typically pass more than one-half of our lifetime in or around our homes. To the extent that our housing improved, it contributed to our level of well-being, happiness, or quality of life.

Unfortunately, as with many fascinating questions, this one is largely intractable. The quality level of a given unit of housing has many dimensions. One may look, for example, at floor area, exterior finish, type of construction, interior detail, presence of amenities such as running water and central heating, extent of repair work needed, or crowdedness. When we renovate or replace an older dwelling, typically some things are gained and others lost. For example, a new dwelling might have a central heating system but lack the 3.5-metre ceilings of the old unit, or might be based on drywall rather than wetwall construction. To say that the new dwelling is better is, therefore, almost always a value judgment. It is an improvement if you think that the characteristics present in new dwellings more than offset the characteristics lost in the older dwelling. It is worse if you think otherwise.

The problem of interpreting quality change is further complicated by two considerations. First, there is a tendency to think of what is modern as being of better quality. This is an insidious viewpoint. When we compare housing stock changes over time, we often tend to note how far the housing stock has come since a certain date. Too often, we then automatically jump from "how far" to "how improved." In some cases, this may be appropriate. I share the belief that inside running water is an important indicator of housing quality. However, other changes in the housing stock may or may not represent an improvement.

Second, censuses include only a limited number of housing quality items. The quality attributes that go unmeasured may be more important than those that are measured. For example, no postwar census inquired into the fire resistance of a dwelling. My personal view is that fire resistance is an important aspect of dwelling quality. To the extent that fire safety was improved over the postwar period, the quality of the housing stock has gone up considerably.

With these two considerations in mind, let us consider the following, more tractable question. What have been some of the important changes in Canada's postwar housing stock? Changes are important in assessing quality improvement.

Exterior Material

In the 1941 and 1951 sample censuses of housing, enumerators recorded the principal material used for the building exterior. These data, which cover all private dwellings, including apartments, are summarized in Table 58. In 1941, about 60 per cent of all dwellings were in buildings with a wood exterior; 30 per cent had a brick or brick veneer exterior, and the remainder had a stucco, imitation siding, stone, or other exterior. Brick exteriors dominated in Ontario and Quebec, especially in Toronto and Montreal. Elsewhere, wood-sheathed buildings formed about 80 per cent of the dwelling stock. Between 1941 and 1951, there

TABLE 58
Occupied dwellings (000s) by exterior facing material, Canada, 1941 and 1951

Exterior facing	*1941*	*1951*
Wood	1,543	1,656
Brick or brick veneer	753	994
Stucco	151	287
Stone	50	62
Imitation siding, other	79	410
All dwellings	2,576	3,409

SOURCE: Taken from the 1941 and 1951 censuses of Canada

was little change in the relative number of brick-faced dwellings or in the predominance of that form in central Canada; the net change in brick-faced dwellings in Toronto and Montreal accounted for almost one-fourth of the national total. However, there was a substantial shift away from wood-sheathed structures. Exteriors of stucco, imitation siding, stone, and other materials rose from 10 per cent to 18 per cent of all dwellings. Only 123,000 wood-sheathed dwellings were added to Canada's housing stock in net between 1941 and 1951. This reflects a low level of construction of new wood-sheathed dwellings and the conversion of older dwellings from wood-sheathed to another exterior facing.

In Canada, there had for some time been a move away from structures built entirely with wood. In part, this was because of cost of maintenance and limited durability. In part, it was also because of the fire hazard. The latter concern prompted some local governments to insist on masonry or other fire-resistant exterior walls. This was especially true in the case of multi-unit buildings. Where bricks were unavailable or expensive, stucco and, later, aluminum siding were relatively inexpensive alternatives. Given the concern with fire-resistant construction, a trend away from wood exteriors is not surprising as Canada's population urbanized.

At the same time, one must be careful in interpreting such data. It is incorrect to equate buildings having an exterior wood facing with those built entirely of wood. As described in Ritchie et al (1967, 171), the "balloon frame" dwelling was introduced into Canada by 1870. It rapidly became the dominant form of construction for single-unit dwellings such as detached, attached, semi-detached, and row houses. In balloon frame construction, stud walls are constructed using 2″ × 4″ or 2″ × 6″ strips of lumber (studs). These stud walls are load-bearing, transporting the weight of each floor and the roof down to the foundations. On the exterior, the stud walls are covered with various materials to provide rigidity and thermal and moisture protection. Finally, the exterior is covered with a finishing material, such as wood, shingles, aluminum siding, or a single layer of brick (called brick veneer). The exterior facing need not be wood, in other words. Another popular construction technique employs masonry walls, typically either concrete block wall or "hollow brick" wall.[9] Masonry walls are load bearing and serve the same function as stud walls. Masonry walls can be left as is or covered with diverse materials ranging from stucco to wood. The exterior facing could be wood, in other words, even though the load-bearing walls are masonry.

We are interested in exterior material used insofar as it indicates change in the quality of the housing stock. Has there been a shift toward more durable, or "better" housing? It is almost impossible to say from these data. This is unfortunate. The postwar period witnessed some substantial changes in residential construction technology.[10] One was the emergence of plywood as a sheathing material, replacing planking. Plywood is typically easier to install and more draught-resistant. Another was the use of prefabricated roof trusses, which are stronger and less expensive than onsite joist constructs. There was also increased use of components constructed at a factory or shop and carried to the site for installation. In general, prefabrication allows for better quality control than is possible in onsite construction. Other important innovations were the use of steel girders in place of wood sills and changes in window design, including double- and triple-glazed windows, that provided improved insulation. Finally, Ritchie et al (1967, 227) mention developments in brick-making technology leading up to the development of the "tunnel kiln" after the Second World War that made large quantities of high-quality bricks readily available for the first time. These technological changes were important improvements in housing quality, but there are virtually no data available about their incidence in the housing stock.

Dwellings in Need of Major Repair

Another important aspect of housing quality is state of repair. A reduction in the number of dilapidated dwellings is one indicator of improvement in the housing stock. Decennial censuses, with the exception of 1971, have included counts of dwellings in need of "major" repair. How useful are these data in measuring quality change?

In assessing whether a dwelling is in need of "major" repair, three kinds of information are needed. First, we must have standards for defining a dwelling that is in "good" repair. These standards are normative; views vary as to what constitutes a dwelling in good repair. Second, we must have information about aspects in which the dwelling falls short of these standards. Sometimes this information can be obtained from casual observation (e.g. sagging foundation or roof), but at other times careful technical analysis is required (e.g. inadequacies with regard to insulation or the water system). Finally, we need to draw a line between minor and major repair. In other words, how far from good repair does a dwelling have to be to need "major" repair?

The principal problem in comparing various census counts is that the approach employed has varied over time. In the 1941 and 1951 censuses, a dwelling was in need of "major repair" if it was seen by the enumerator to possess any of the following defects: "(a) sagging or rotting foundations, indicated by cracked or leaning walls; (b) faulty roof or chimney; (c) unsafe outside steps or stairways; (d) interior badly in need of repair - that is large chunks of plaster missing from walls or ceiling" (*1951 Census of Canada*, vol 10, 373). In the 1961 Census, the concept of a major repair was somewhat different: "In the 1951 Census, broken or unsafe outside steps or stairs were considered to be an indication of a need for major repair, while in 1961 . . . the emphasis [was] put on sagging or crumbling foundations, faulty roof or chimney, rotting doorsills and window frames and major signs of deterioration inside the building" (*1961 Census of Canada*, vol 7[2], 4.24). In the 1981 Census, respondents were asked to indicate whether their dwelling was in need of repair, excluding desirable remodelling or additions. Possible responses included "needs only regular maintenance," "needs minor repairs," and "needs major repairs." The following clarification was supplied by Statistics Canada: "Regular maintenance refers to painting, furnace cleaning, etc. Minor repairs refers to missing or loose floor tiles, bricks, or shingles, defective steps, railing or siding, etc. Major repairs refers to defective plumbing or electrical wiring, structural repairs to walls, floors or ceilings, etc" (*1981 Census Dictionary*, 82-3).

Table 59 shows the percentage of dwellings found to be in need of major repair in each census. Available breakdowns by geographic locale, tenure, and type of dwelling are also shown. The 1941 and 1951 data are comparable. Comparisons are also possible between 1951 and 1961, although allowance must be made for the dropping of broken or unsafe steps as a major repair item. Comparisons between 1961 and 1981 are tenuous because of the substantial change in measurement method.

These data indicate substantial reduction between 1941 and 1961 in the incidence of dwellings in need of repair. This reduction occurred in both urban and rural areas, in the owner-occupier and rental sectors, and in both single detached and other housing forms. The 1981 data show a somewhat higher incidence of houses in need of major repair but could result from the change in approach to measurement.

I am sceptical about these data on several accounts. First, the standards for a house to be in good repair are ill-defined. The 1941, 1951, and 1961 censuses employ at most four (three in the case of 1961) criteria as

TABLE 59
Occupied dwellings showing percentage in need of major repair by geographic area, tenure, and dwelling type, Canada, 1941–81

	1941	*1951*	*1961*	*1981*
All dwellings	27	13	6	7
Rural	36	17	9	–
Urban	20	8	4	–
Owned	–	13	5	–
Rented	–	15	7	–
Single detached	–	14	6	7
Owned	–	13	5	–
Rented		19	12	–
Other dwellings	–	12	5	6

SOURCE: Computed from published census reports, various years

outlined above. The 1981 Census criteria are fuzzy and subject to a variety of interpretations. Second, the censuses do not undertake a thorough evaluation of structural problems in the dwelling. In the 1941–61 censuses, the enumerator undertook a quick visual inspection and could not have known of any "hidden" major repair items. The 1981 Census allowed the respondent wider grounds in deciding whether the dwelling was in need of major repair but provided no systematic grounds for such evaluation. Third, there is no explicit description in any of the censuses of the dividing line between "major" and "minor" repairs.

Heating System

Some remarkable changes in the heating systems employed in Canadian dwellings are evidenced in Table 60. The principal change was the rise of the central heating system, which includes furnaces and installed electric heating. In 1941, over 60 per cent of all dwellings were still heated principally by a stove. Although primarily a rural phenomenon, stoves were the principal source of heating in 44 per cent of urban dwellings as well. By 1951, the figure had dropped to 50 per cent of all dwellings (37 per cent of urban homes). In the 1961 Census, stoves are lumped together with space heaters, but together they still accounted for just 29 per cent of all dwellings. The large absolute decline in stove-heated homes in the 1950s came about in part because of the extensive conversion of such homes to central heating. Through the 1960s, new dwellings and converted existing dwellings were equipped mainly with furnaces.

Related to this have been changes in principal heating fuels. See Table 61. In 1941, almost all dwellings were heated by coal, coke, or wood. By 1981, the use of these heating fuels was negligible. During the 1950s and

TABLE 60
Occupied dwellings (000s) showing type of principal heating equipment, Canada, 1941-81

Principal heating equipment	*1941*	*1951*	*1961*	*1971*	*1981*
Furnace: steam or hot water	367	529	830	1,337	1,678
Furnace: hot air	630	1,053	2,242	3,202	4,065
Stove or space heater*	1,578	1,724	1,324	1,042	552
Installed electric†	-	-	-	353	1,771
Other, unspecified	-	104	158	97	216
All dwellings	2,576	3,409	4,554	6,031	8,282

SOURCE: Taken from published census reports, various years
* 1941 and 1951 census data do not include space heaters with stoves.
† Installed electric included with other prior to 1971

1960s, liquid fuels and natural gas and propane became the dominant fuels. In large part, this was because hard fuels (coal, coke, and wood) required extensive handling and were not suited to the automated central heating systems that became predominant. In part, this also reflects the abundance of liquid fuels and gas and their low cost. Electricity began to become important during this period, too, but it became much more popular in the 1970s. During the 1970s, oil began to wane as a heating fuel, reflecting its high price, sometimes limited availability, and the active promotion of "off oil" conversion programs by federal and provincial governments.

What did these changes in heating equipment and fuel imply about the quality of the housing stock? In part, this is a matter of personal preference. There are those who miss the cheery warmness of a wood-fired kitchen stove or the quiet pleasures of stoking a coal furnace. Others argue that the convenience of an automated central heating system represents an important quality improvement and that the switch away from coal and wood as fuels has also improved life in urban areas by reducing pollution.

Electrification

Virtually all the appliances that helped to make postwar dwellings safer and more comfortable were electrically powered. Thus, electrification was an essential antecedent. Further, as the variety and number of these appliances increased, successive changes to the dwelling's power distribution system were necessitated. For example, the amperage of the typical residential entry panel increased over the years to accommodate larger power demands. Another example was the redesigning of circuits, especially in kitchens, to reduce the likelihood of overloading when several appliances were used simultaneously. Also, several changes were

TABLE 61
Occupied dwellings (000s) by type of principal heating fuel, Canada, 1941–81

Principal heating fuel	*1941*	*1951*	*1961*	*1971*	*1981*
Coal or coke	1,207	1,470	481	66	*
Wood	1,184	951	590	213	*
Oil, liquid fuels†	67	774	2,565	3,441	2,802
Gas	87	163	858	1,936	3,132
Electricity	6	8	31	353	2,005
Other, not stated	26	43	30	23	342
All dwellings	2,576	3,409	4,554	6,031	8,282

SOURCE: Taken from published census reports, various years
* Coal, coke, and wood fuels included in "Other, not stated" in 1981
† 1941 Census provides separate data only for fuel oil.

made to promote safety, e.g. the increasing use of three-prong outlets. Many such changes were mandated by law and automatically integrated into new dwellings. Over the years, some of the older housing stock was also upgraded. These changes represent a major improvement in the housing stock. Unfortunately, there are no comprehensive data on the extent of such improvements.

Early censuses do report on the number of households with electric lighting. By 1941, almost all urban households had become electrified. However, electrification came later to rural areas. In 1941, just 20 per cent of rural dwellings had electric lighting. Further, many rural areas and smaller towns had only a 25-hz supply. The conversion of the rural housing stock to electric power, and the switch to 60-hz, were important developments in the early postwar period. Censuses did not report on the upgrade from 25- to 60-hz. However, the 1951 Census reports that almost two-thirds of rural dwellings had become electrified. The subsequent widespread use of electricity contributed to a decision to drop this question from the 1961 Census.

The rapid spread of appliances that followed electrification is well known. In many cases, the diffusion was so fast that an appliance, too rare to count in one census, was too common to be worth counting in the next. Succeeding censuses attempted to count some of the principal, but not yet universal, new appliances of the time. Table 62 shows the number of occupied dwellings in Canada having a specified appliance in postwar censuses.

Does the growing use of such appliances represent an improvement in the housing stock? In part, this is a definitional problem in which we are asking what we mean by housing. One has to draw a line between housing and other goods. Do we want to include certain appliances as part of "housing?" Table 62 serves, if nothing else, to illustrate the importance

TABLE 62
Occupied dwellings (000s) presence of appliances, Canada, 1941–71

Appliance	*1941*	*1951*	*1961*	*1971*
Clothes dryer	–	–	–	2,429
Electric dishwasher	–	–	–	789
Electric vacuum	624	1,414	–	–
Home freezer	–	–	678	2,022
Mechanical refrigerator	539	1,595	4,145	5,915
Radio	2,003	3,144	–	–
Television	–	–	3,757	5,750
All dwellings	2,576	3,409	4,554	6,031

SOURCE: Taken from published census reports, various years

of residential electrification (and upgrading) in terms of the uses made of it.

In chapter one, it is argued that one reason for the rapid postwar formation of households was the improvement in home-making technology. The appliances described above, among others, have made it easier for small households, and particularly persons living alone, to manage their eating and cleaning activities. The data presented above, while fragmentary, indicate that the acquisition of such appliances has been rapid and widespread, providing some support for the argument.

Water, Bath, and Toilet Facilities

There were also some major changes in the housing stock in terms of water, bath, and toilet facilities. By 1971, almost all dwellings had piped hot and cold running water, an installed bath, and an inside flush toilet, a substantial change in just three decades.

In 1941, only about 60 per cent of dwellings had piped running water. Almost one dwelling in four relied on an outside source, be it a hand pump outside the dwelling or a nearby lake. Those dwellings without inside running water were preponderantly rural. However, even among urban dwellings, almost 10 per cent were without piped water in 1941. As shown in Table 63, there was a marked improvement during the 1940s; by 1951, almost three-fourths of all dwellings had piped running water. In part, this was because of rapid urbanization. Also, within rural areas the incidence of piped water rose from under 25 per cent in 1941 to almost 40 per cent in 1951; in urban areas, it rose from 91 per cent to 95 per cent. Similar increases were experienced in the 1950s and 1960s. By 1971, 96 per cent of all dwellings had piped water. Virtually all the housing stock added during the postwar period was equipped with piped

TABLE 63
Occupied dwellings (000s) by water, bath, and toilet facilities, Canada, 1941-71

	1941	*1951*	*1961*	*1971*
Total occupied dwellings	2,576	3,409	4,555	6,035
Water supply				
Piped running water*	1,559	2,524	4,058	5,794
Hot and cold	-	1,940	3,650	5,593
Cold only	-	584	408	202
Other source	1,018	886	496	240
Inside pump, private source	247	-	-	-
Outside supply	771	-	-	-
Bath or shower				
Private installed	1,170	1,938	3,512	5,486
Shared installed	100	135	147	78
None	1,305	1,336	895	450
Toilet facilities				
With inside flush toilet	1,450	2,329	3,881	5,702
Private	1,342	2,187	3,599	5,627
Shared	108	142	281	75
Without inside flush	1,126	1,081	674	332
Chemical toilet	41	82	62	-
Outside privy, other	1,085	999	612	-
Connection to sewer	-	-	2,903	4,432
Septic tank, other	-	-	1,652	1,598

SOURCE: Taken from published census reports, various years
* In the censuses, "piped water" was defined as water the flow of which could be controlled by means of tap. Water from a hand pump located within the dwelling was not considered as piped running water.

water. In addition, these figures suggest a massive conversion of the older housing stock. In some cases, this was a conversion from a private well or other source to a municipal water line. In others, the conversion was from hand pump to electric pump.

For many dwellings, hot running water came later. In the 1951 Census (there are no corresponding data for 1941), 60 per cent of occupied dwellings had hot running water. The remainder presumably used stoves or other devices to heat water as required. Again, the absence was most marked in rural dwellings. However, even in urban areas, almost one household in four lacked hot water. The rapid increase in dwellings with hot running water during the 1950s and 1960s suggests extensive conversions of older dwellings to provide hot water. In part, this reflects the postwar rural electrification program that made electric water heaters feasible for many households for the first time.

The trend toward installed baths and showers matched that for hot water heating. In the absence of piped hot water, bath water was typically heated on a stove. The bathtub could be, and often was, portable. It was moved to a room, and hot water was carried to it. However, with piped hot water, it became possible to deliver water directly to the tub. This necessarily fixed the position of the bathtub, leading to the emergence of the installed tub. Also, piped hot water made (pleasant) showers feasible for the first time.

The trend toward inside flush toilets similarly matched that for piped running water. Such toilets typically use over 20 litres of water per flush. Refilling flush toilets in the absence of piped running water is onerous, to say the least. Typically, one would instead use some other form such as a chemical toilet or an outside privy. However, given the availability of piped water, the conveniences of a flush toilet proved attractive to most households.

The censuses of 1961 and 1971 also provide information on methods of sewage disposal. These show an increase in the proportion of dwellings connected to a sewer system, usually municipally operated. The remainder use a variety of disposal methods including septic tanks, cesspools, and pits. Again, the difference between rural and urban dwellings is large, reflecting the infeasibility of municipal sewer connections for many rural dwellings.

The data in Table 63 indicate a major quality improvement in the housing stock. In part, this has been brought about by the construction of new dwellings which include these amenities. In part, it represents a substantial refitting of the older housing stock with modern plumbing and toilet facilities.

Age of Dwelling

Another commonly used indicator of housing quality is age of dwelling. In general, dwellings deteriorate with the passing of time. Roofs wear out; foundations crumble; wood sills and frames rot. This is a kind of decline in quality. Economists often characterize a housing unit as providing a certain "flow of housing services." The flow of housing services provided by a dwelling is argued to decline as the dwelling deteriorates. This is an elegant if abstract way of describing reductions in dwelling quality with aging.

Entering the postwar period, Canada's housing stock was relatively old. Early postwar estimates in Firestone (1951, 44-9) suggest that, at the end of 1949, the average age of a Canadian house was 33 years. Further, it is suggested that the average age had increased steadily since at least 1921 (when it stood at 21 years). Why? It is not difficult to see how the lack of construction during the Depression contributed to an aging housing stock. However, Firestone thinks that the housing stock aged even during the 1920s, which were relatively prosperous. To explain this, one must look at the immigration into Canada that occurred between 1901 and 1911. During this decade, Canada's population increased more rapidly (3.0 per cent annually) than at any other point in its history. Because this population growth was driven by immigration, it translated almost immediately into household formation and housing demand. Massive new construction meant a reduction in the average age of a dwelling; however, I am not aware of any empirical estimates.[11] Thus, by 1921, the housing boom induced by this population growth had run its course. The 1920s were a period of lower population growth (2.0 per cent annually), with very little immigration. As a consequence, there was relatively little new residential construction, and the age of a typical dwelling increased.

The postwar period was characterized by considerable reduction in age of dwelling. In the 1950s, the volume of new dwelling construction began to climb sharply. Firestone (1951, 267) estimates that annual new completions rarely exceeded 60,000 units in the 1920s and 50,000 in the 1930s. As seen in Table 47 above, completions were much higher in the 1950s, ranging from 73,000 to 147,000 dwellings. By 1961, 44 per cent of all occupied dwellings had been constructed within the preceding 15 years. See Table 64. Through the 1960s and 1970s, large volumes of new construction continued to mean declines in the average age of dwelling. By 1981, more than one-half of all occupied dwellings were less than 20 years old.

Does this postwar reduction in the average age of a dwelling imply an improvement in the quality of the housing stock? In certain respects, one can answer affirmatively. To the extent that new housing integrated many of the emerging amenities such as modern plumbing, heating, and electrical systems, I think it has. Also, to the extent that badly deteriorated dwellings have been replaced by sound construction, one can also be affirmative. Commenting on the aging of the housing stock up to 1949, Firestone (1951, 49) states: "In some measure this aging process reflects deterioration in the quality of living accommodation. This factor should be borne in mind in any considerations whether the Canadian standard of housing accommodation is higher today than it was at any preceding period."

However, there is not necessarily a simple connection between building age and deterioration. By undertaking maintenance and renovation, one can keep at least some older buildings in relatively good shape. Sometimes, of course, the costs of maintenance and renovation are too high. At other times, though, maintenance and renovation can be both feasible and efficient. Also, we must remember that older housing was constructed in different ways from newer housing. There are differences in material, in design and layout, and in the details of construction and

TABLE 64
Occupied dwellings (000s) by period of construction, Canada, 1961–81

Period of construction	*1961*	*1971*	*1981*
Total occupied dwellings*	4,555	6,031	8,282
1919 or earlier	1,392	–	–
1920 or earlier		1,202	885
1920–45	1,148	–	–
1921–45	–	1,089	1,053
1946–59	1,846	–	–
1946–60	–	2,000	1,856
1960–61†	168	–	–
1961–65	–	807	–
1966–69	–	711	–
1970–71†	–	221	–
1961–70	–	–	1,800
1971–75	–	–	1,325
1976–79	–	–	1,109
1980–81†	–	–	253

SOURCE: Taken from published census reports, various years
† Includes only dwellings completed and occupied by 1 June of the census year
* These data are taken from the sample census. Because they are "blown up" estimates from a 20 per cent (1961, 1981) or 33 percent (1971) sample, total counts may vary from counts based on the 100 percent questionnaire.

finishing. Some of these are described above. To some people, these differences are important; they prefer the features found in older dwellings. Unfortunately, censuses did not inquire into such building details. Therefore we have no systematic source of information on these desirable qualities of older dwellings, qualities that may well be rapidly disappearing.

CONCLUSIONS

This chapter is descriptive. It details the size and character of Canada's housing stock and its changing nature over the postwar period. After chapter four, where the rapid growth of households is noted, and remembering the equivalence of households and occupied dwellings, the rapid growth of the housing stock described in this chapter should not be surprising.

What is perhaps more surprising, at least to the unitiated, are the many different changes that have occurred in the characteristics of the housing stock. The decline of the single detached dwelling and its later renaissance, the rise, fall, and then re-rise of owner-occupancy, the surprisingly constant average size of dwelling, and the rapidly changing quality attributes of the housing stock are all noteworthy. Much of this change has been accommodated at the margin, i.e. by new dwellings with certain characteristics replacing older dwellings that were different. However, in many cases, the changes were too large or abrupt to be simply that; a widespread conversion, or retrofitting, of older dwellings was also responsible.

Unfortunately, we do not have a clear picture of the extent of conversion activity. We do not know just how malleable the older housing was in particular respects. Early in the chapter, an argument is made for an improved system of housing accounts data. Such an accounting scheme would, among other things, permit an assessment of the extent and nature of conversion activity. With such data, some of the questions raised in this chapter could be addressed.

In this chapter, I have also speculated on some of the reasons for the observed changes in the housing stock. Some of these are what might be called supply considerations and include changes in costs and building technologies. Others are demand considerations; these include demographic changes and changes in income. Finally, some are related to housing policies of the various levels of government. We return to some of these in the two ensuing chapters, using other data sources. My purpose in mentioning them first in this chapter is merely to illustrate some possible explanations in a preliminary way and to show the limited usefulness of published census reports in exploring them.

CHAPTER EIGHT

Prosperity and Housing Affordability

The preceding chapter examines how Canada's housing stock changed after 1945. It is essentially descriptive. It depicts, but does not extensively explain, the changes that occurred. Chapter nine argues that the housing stock changed in part because of federal, provincial, and local housing policies. Although housing is a good predominantly traded within the private sector, governments had a substantial influence on its provision. Nonetheless, it is instructive to consider housing as simply another good provided in a private market. An economist envisages a demand curve for housing (showing how much housing consumers are willing to purchase, depending on price and other demand-related variables) and a supply curve (showing how much housing is put onto the market, depending on price and other supply-related variables). In a market in equilibrium, changes in the housing stock can then be traced to changes in the demand and supply curves.

One cause commonly used to explain postwar changes in the housing stock is rising prosperity. Canadians used their increasing affluence in part to purchase more and better housing. This is a demand-based argument. With rising incomes, the demand curve shifted to the right and there was a consequent increase in the stock of housing. At the same time, other factors reshaped the demand for housing. Demographic shifts - in the age, size, and compositional mix of households - were important, for instance. In addition, there were changes to the supply curve; new construction technologies, building materials, and methods of financing new housing, for example, also contributed to housing stock change.

This chapter examines the link between rising incomes and housing stock change. We shall look at changes in income relative to the price of housing. As incomes rose relative to shelter costs, housing became more affordable, and consumption increased. Comparing incomes to shelter prices, i.e. emphasizing housing affordability, is also useful in assessing other pertinent demand and supply factors. For example, where new construction techniques improve efficiency, housing prices may fall either absolutely or relative to incomes. A drop in price means that consumers are better off. Even if their incomes do not rise, they are able to purchase more housing, or goods of other kinds. Thus housing affordability is an indicator of overall prosperity. While this is also true of other commodities, the relatively large budget share devoted to housing makes it especially important for most consumers.

Just how consumers responded to the changing affordability of housing is the focus of this chapter. The price sensitivity of demand depends on the availability of substitutes. If the price of a good increases, consumers attempt to substitute a less expensive alternative. What might that alternative be? The consumer might substitute another kind (e.g. size, type, tenure) of dwelling that is nearly as desirable. Alternatively, the substitute may be another good altogether, for example, foresaking a larger house to live in a smaller unit while taking more vacations. Another possibility is that the consumer (i.e. the persons making up the consuming unit) may either disband or not form. For example, a young woman might want her own apartment but instead choose to remain in her parents' home if rents are too high. The price sensitivity of demand will reflect the feasibility and desirability of such substitutions.

This chapter starts with a discussion of how housing consumption changed between 1945 and 1981. Next, we look at how to measure the price of housing. We assess postwar trends in several housing price indicators. Some indices of prosperity are introduced, and postwar changes described, focusing on the significance of the working wife, the lone parent, and the unattached individual. Then some aggregate models of postwar housing demand are reviewed. In these models, the effects of demographic change are either ignored or treated crudely. In effect, single estimates are made of price and income elasticity. Disaggregate models are subsequently considered in which housing demands are estimated separately for different kinds of demographic groups. This allows us to look at the joint effects of changing household composition, rising incomes, and changing prices on the demand for housing.

POSTWAR HOUSING CONSUMPTION

Chapter seven considers growth and change in the postwar Canadian housing stock. Overall, the number of dwellings increased rapidly. The quality of the stock also improved in several respects, in part because of the higher standards used in new construction and in part because of a retrofitting of the existing stock. Looking at changes in the number of

TABLE 65
Gross national expenditures ($ billions) and selected components, Canada, 1946–81

Year	Gross national expenditure	Gross rent, fuel, and power	Gross fixed capital formation	
			Total	Residential
1946	11.9	1.2	1.7	0.4
1951	21.6	1.5	4.4	0.8
1956	32.1	3.3	8.0	1.8
1961	39.6	4.8	8.4	1.8
1966	61.8	6.6	15.4	2.6
1971	94.5	10.6	20.8	4.8
1976	191.9	20.0	44.9	12.4
1981	339.8	38.2	82.1	16.4

SOURCES
Statistics Canada. 1976. *National Income and Expenditure Accounts, Vol 1, 1926–1974.* Catalogue No. 13-531. Tables 2, 53, and 53A
– 1985. *National Income and Expenditure Accounts, 1970–1984.* Catalogue 13-201. Tables 2 and 53

dwellings or in the number with a specific quality characteristic is one way of describing the increased level of housing consumption.

Another way of describing these changes is in terms of increased expenditure on housing. What is remarkable about the growth in expenditure is its consistency. See Table 65. Between 1946 and 1981, gross national expenditure (GNE) increased by about 30-fold in current dollars. Consumer expenditure on gross rent, fuel, and power (which includes both the out-of-pocket costs of shelter plus imputed rent on owned homes), as a component of it, kept pace at about 10 per cent of GNE. Investment in residential fixed capital formation, though somewhat more volatile, tended to make up about 25 per cent of total gross investment.

This steady growth in consumption and construction is surprising in light of the rapid shifts in household formation and the postwar surge in real incomes. With the shift to smaller households, for example, one might have anticipated new patterns of housing consumption. In addition, there was no particular reason for consumers to maintain a constant growth in housing consumption as they became more prosperous. Didn't rising incomes also result in new patterns of consumption (i.e. larger budget shares for luxury goods, smaller shares for other goods)?

The steady growth in consumption is also surprising in view of the changes in tenure and dwelling type. In chapter seven, it is argued that tenure patterns changed only modestly overall. However, the differences are much larger when disaggregated by size of household. See Table 66. Among persons living alone, 44 per cent were home owners in 1941, com-

TABLE 66
Percentage distribution of households by tenure and (for renters) by dwelling type, by size of household, Canada, 1941 and 1981

	Persons in household				
	One	*Two*	*Three*	*Four*	*Five +*
(a) 1941					
All households	100	100	100	100	100
Home owners	44	43	44	46	47
Renters					
Single	23	20	23	24	24
Semi-detached, row	4	7	8	8	9
Apartment, flat, other	28	30	25	22	20
(b) 1981					
All households	100	100	100	100	100
Home owers	32	59	68	79	82
Renters					
Single	6	6	7	6	7
Semi-detached, row	4	5	6	5	5
Apartment, flat, other	57	31	19	10	6

SOURCE: Calculated from published reports of the 1941 and 1981 censuses

pared to just 32 per cent in 1981. Among larger households, the home ownership rate rose (e.g. from 47 per cent of households of five or more in 1941 to 82 per cent in 1981). At the same time, there was a big shift on the part of smaller renter households into apartments. Of course, changes in tenure or type of dwelling need not necessarily imply changes in expenditure. However, the contrast between these changes and the steady growth of expenditure is interesting.

Looking at expenditures on "gross rent, fuel, and power" is just one way of characterizing housing expenditure. What is spent on housing depends partly on how one separates housing from other consumption. In the Family Expenditure Surveys that Statistics Canada has undertaken periodically since 1947, a different definition is employed. It is broader in that it includes property taxes, repairs, mortgage interest, insurance premiums, furniture, appliances, floor covering, some household textile products, utensils, and household equipment, supplies, and services. However, it does not include the imputed rent on home owner equity. Statistics Canada uses these surveys to derive weights for housing and other goods in determining the consumer price index. The weights for housing, shown in Table 67, are divided into two groups, one for shelter, the other for household operation. The overall housing expenditure weight remained roughly constant for much of the postwar period, increasing modestly only in the 1970s. However, some housing expenditure components became more important, others less important.

TABLE 67

Housing expenditure weights in the consumer price index, Canada, 1949-78

Component	*1949*	*1957*	*1967*	*1974*	*1978*
Total housing expenditure	32.0	32.2	31.4	34.1	35.4
Shelter expenditure	15.0	17.9	17.9	19.1	20.7
Renting					
Rentals	8.6	8.6	8.3	6.6	7.0
Tenant repairs		0.2	0.1	0.1	0.1
Tenant insurance premiums				0.0	0.0
Home ownership					
Property taxes	1.5	2.2	2.9	2.7	2.7
Insurance	0.3	0.2	0.3	0.3	0.6
Repairs	1.4	1.4	1.0	1.9	1.9
Mortgage interest	0.9	2.4	3.1	3.3	4.7
Replacement	2.4	2.9	2.3	3.0	2.7
Other owned accommodation expense				0.8	0.5
All other accommodation expense				0.4	0.5
Household operation expenditure	17.0	14.3	13.5	15.0	14.6
Fuel and lighting*	4.8	4.4	3.8	3.3	3.8
Furniture†	2.2	1.9	1.7	2.2	2.0
Appliances‡	1.9	2.2	1.4	1.4	1.2
Floor coverings§	0.6	0.5	0.6	0.8	0.2
Textiles#	0.9	0.7	0.6	0.7	0.6
Utensils and equipment‖	1.5	1.0	0.7	1.2	1.4
Household supplies**	2.1	1.5	2.0	2.4	2.4
Household services††	3.0	2.1	2.7	2.9	3.0

SOURCES

Statistics Canada. 1952. *The Consumer Price Index: January 1949–August 1952.* Catalogue 62-502

–1961. *The Consumer Price Index for Canada (1949 = 100).* (Revision based on 1957 expenditures.) Catalogue 62-518

–1973. *The Consumer Price Index for Canada (1961 = 100).* (Revision based on 1967 expenditures.) Catalogue 62-539

–1978. *The Consumer Price Index for Canada.* (Revision based on 1974 expenditures.) Catalogue 62-546

–1982. *The Consumer Price Index Reference Paper.* (Concepts and Procedures: updating based on 1978 expenditures.) Catalogue 62-546

* Includes coal, gas, fuel oil, and electricity. Coal dropped in 1967. Water added in 1974

† Living room suites, bedroom suites, dinette suites, kitchen table, chairs and mattresses listed in 1949. In 1967, description changed from suites to furniture. Box springs added in 1974; also definitions of living room furniture further elaborated

‡ Refrigerator, stove, washing machine, vacuum, iron listed in 1949. Automatic dryers, electric frying pan, sewing machine added in 1957. In 1967, electric frying pan and iron not listed; new category, small appliances, added. In 1974, dishwashers, electric food mixers and blenders added. In 1978, description changed for minor appliances to include cooking, food warming and preparation, other small appliances

(continued)

TABLE 67 (continued)

§ Carpet and linoleum listed in 1949. In 1967, latter replaced by hard-surface floor coverings. Restricted to area rugs and mats in 1978

Cotton sheets, wool blankets, towel, drapery, plastic listed in 1949. In 1967, list altered to bed sheets, bed spreads, towels, draperies. In 1974 slipcovers, curtains, and related yard goods added. Pillowcases, pillows, blankets, washcloths, and dishcloths added in 1978

|| 1949 list includes dishes, glassware, glass ovenware, enamel or aluminum saucepans, broom, garbage can, alarm clock, light bulb, and lawnmower. Enamel saucepan, broom deleted in 1957; hardware added. References to glassware, glass ovenware, aluminum saucepan, garbage can, alarm clock, light bulb, hardware, and lawnmower dropped in 1967; replaced by flatware, cookware, refuse containers, and house and gardening tools. In 1978, other metal flatware and serving pieces, earthenware, and bone china added. Since 1974, list includes metal or glass cookware, stainless steel flatware, dinnerware, garden and power tools, other tools, outdoor equipment, refuse containers

** 1949 list includes ice, soap, bleach, scouring powder, floor polish, toilet paper, waxed paper. Ice deleted in 1957; detergent added. Fabric softener, stationery, garbage bags, garden supplies, and flowers added in 1967; waxed paper replaced by food wrap. In 1974, list rewritten to include furniture polishes, pet care, facial tissue, kitchen paper products, plastic garbage bags, plastic food bags and wrap, foil stationery, garden supplies, flowers and light bulbs. Dry cell batteries and flourescent tubes added in 1978

†† 1949 list includes telephone, laundry, dry cleaning, postage, household help, household effects insurance, shoe repairs. Laundry, dry cleaning, shoe repairs deleted in 1957. Appliance repairs added in 1967. Since 1974, list includes child care, domestic help, servicing and repair of equipment and furnishings, local and long-distance telephone, postage.

The Family Expenditure Surveys also provide useful information on the shelter expenditures of different kinds of consumers over time. In Table 68 are presented comparative data on the percentage of total expenditure on housing (i.e. excluding household operation expenses). The relative stability of shelter expenditure is again evident. This is in sharp contrast to the substantial differences in shelter expenditure between different kinds of households. The elderly, especially those living alone, allocate a larger portion of their total consumption to shelter. Middle-aged families spend relatively much less.

Table 68 reconfirms the quandary mentioned at the outset. Given the rising incidence of living alone and increasing number of elderly, and their higher propensity to spend on housing, why didn't the overall growth in shelter expenditure outstrip other expenditures in postwar Canada? I think the answer lay in part in the price and income elasticities of housing demand.

TABLE 68
Shelter expenditure as a percentage of total expenditure by size of spending unit and age of head, Canada, 1947-82

Year	All spending units	One-person unit				Two persons or more			
		All	Under 45	45-64	65 or older	All	Under 45	45-64	65 or older
1947-8*	15	18	-	-	-	14	-	-	-
1959	17	22	-	-	-	16	17	15	20
1964†	17	22	18	20	28	16	17	15	20
1967	16	21	15	25	32	16	16	14	18
1969†	16	22	14	20	28	15	-	-	-
1978	16	21	18	22	29	16	17	13	20
1982	18	22	19	20	30	17	-	-	-

SOURCE: Calculated from *Family Expenditure Survey* (various years) published by Statistics Canada

* 1947-8, 1959: Sample of units drawn from cities of 15,000 population or more. Includes both families and unattached individuals

† 1964, 1967: Sample of units drawn from 11 large cities in Canada. Includes both familie-sand unattached individuals

‡ 1969, 1978, 1982: Sample of units drawn from across Canada. Includes rural and urban spending units and both families and unattached individuals

POSTWAR CHANGES IN HOUSING PRICES

One widely used price indicator is Statistics Canada's housing index (SCHI). SCHI is a component of the consumer price index (CPI). See Table 69. Between 1946 and 1981, SCHI increased just under five-fold, about the same pace as the overall CPI. In other words, shelter prices went up at about the same rate as other consumer prices. However, per capita disposable income went up 12-fold. Housing became more affordable to the typical consumer. Indeed, the same could be said of other consumer goods in general.

However affordable housing was becoming, consumers continued to spend a large portion of their budget on housing. See Table 67. The typical budget share remained constant at about 32 per cent until about 1970 and rose moderately thereafter. Thus, over this period, consumers spent substantially more on housing. Even before the 1970s, housing expenditures kept pace with rapidly rising incomes, although shelter prices were rising less quickly. Why?

Two answers come to mind. First, the demand for housing may be income elastic. In other words, consumers may have allocated a large proportion of the growth in their incomes to housing. Thus, even if the

TABLE 69
Housing prices and incomes (1971 = 100), Canada, 1946-81

Year	SCHI*	CPI†	PDIP‡	AUY§	Year	SCHI	CPI	PDIP	AUY
1946	46.4	45.0	26.2	-	1966	79.5	83.5	71.7	-
1947	49.8	49.2	28.6	-	1967	82.9	86.5	76.1	73.7
1948	53.6	56.3	32.4	-	1968	86.7	90.0	81.4	-
1949	54.9	58.0	33.1	-	1969	91.2	94.1	87.2	86.9
1950	57.2	59.7	34.9	-	1970	95.7	97.2	91.3	-
1951	62.4	66.0	39.6	34.4	1971	100.0	100.0	100.0	100.0
1952	64.8	67.6	42.1	-	1972	104.7	104.8	112.4	107.7
1953	65.9	67.0	42.9	-	1973	111.4	112.7	130.1	120.9
1954	66.7	67.4	42.1	39.5	1974	121.1	125.0	152.1	140.6
1955	67.2	67.5	44.3	-	1975	133.2	138.5	176.0	156.1
1956	68.1	68.5	47.7	-	1976	148.1	148.9	196.1	182.0
1957	69.5	70.7	49.2	46.1	1977	161.9	160.8	214.0	189.5
1958	70.8	72.6	51.2	-	1978	174.1	175.2	239.2	209.7
1959	72.1	73.4	52.4	48.8	1979	186.2	191.2	267.5	228.0
1960	72.8	74.3	53.5	-	1980	201.4	210.6	300.0	255.2
1961	73.1	75.0	53.1	52.0	1981	226.4	236.9	343.6	289.9
1962	74.0	75.9	56.8	-					
1963	74.8	77.2	59.2	-					
1964	76.0	78.6	61.6	-					
1965	77.3	80.5	66.4	65.3					

SOURCES
SCHI and CPI taken from *Historical Statistics of Canada, 2nd Edition* (1983, K8-K18) and Statistics Canada (1978, 1981), *Prices and Price Indexes*. PDIP calculated from selected years of *Canadian Statistical Review, System of National Accounts*, and *Estimates of Population for Canada and the Provinces*, all published by Statistics Canada. AUY calculated from *Incomes of Nonfarm Families and Individuals in Canada: Selected Years, 1951-1965* and *Income Distributions by Size in Canada*, both published by Statistics Canada.
* Statistics Canada's housing price index
† Statistics Canada's consumer price index
‡ Per capita disposable income
§ Average unit income (economic families and unattached individuals)

price of housing did not keep pace with income, expenditures did. It is helpful to think of a consumer's housing expenditure (e) as the product of the price of a unit of housing (p) and the amount of housing (h) consumed. What is meant by the amount of housing consumed (h)? Let us think of it as a "flow of housing services," i.e. in terms of the quality of the dwelling.

If the demand for quality is income elastic, expenditure may have risen in step with income because consumers were improving the quality of their housing, even though the price of a standard dwelling increased less quickly. To assess this argument, however, we have to decompose expenditure into two unobservable components: quality (h) and price (p) per unit of quality. How might this be done? Does SCHI or any other index necessarily approximate a unit quality price?

How does Statistics Canada calculate its price index? SCHI is a component of the CPI. The CPI measures changes in the cost of purchasing a fixed bundle of goods: including housing. To fix ideas, suppose the bundle consists of I different goods, and that q_i and p_i are the budget share (at a base date s) and price of the ith good respectively. We can further subscript prices by dates s and t over which the aggregate price change is to be measured. The price index for date t (relative to date s) is:

$$\text{CPI} = \sum_{i=1}^{I} p_{it}q_i / \sum_{i=1}^{I} p_{is}q_i. \qquad (23)$$

The CPI for date s (the base date) is 1.0 by definition. Sub-indices are also available for components of this bundle, e.g. various kinds of shelter expense. A breakdown of the CPI into its housing components is shown in Table 67 above, and Table 70 displays indices for shelter components. From time to time, the bundle is altered to reflect changes in consumption patterns. The reweighting is based on results of the Family Expenditure Surveys, hereinafter referred to simply as the "base surveys."[1]

Such price indices control for quality change in two ways: by holding the bundle constant, and by ensuring that the p_i at different dates stands for exactly the same component. However, because the bundle changes with each new base survey, price index comparisons that span two or more base surveys can introduce quality change. As an example, the budget share of repairs in the shelter price index for owner-occupiers was 22 per cent in 1949, 15 per cent in 1957, and just 11 per cent in 1967. In part, this reflects the declining average age of the housing stock with the extensive new construction of the 1950s and 1960s. Because of this, SCHI became less sensitive to the price of repairs. But this introduces a quality change (dwelling newness) into calculation of the price index.

To examine SCHI more closely, let us divide it into three components: owned shelter expense, rented shelter expense, and household operation.[2] The most troublesome parts are the two tenure expenses.

Shelter Price Index for Renters

Consider first the rent component of the CPI. The method of calculation changed slightly over time. In later years, the following approach was used.[3]

> The data on rents used in the "Rented accommodation" component of the CPI are collected through . . . a sample of some 56,000 households . . . Once a house-

TABLE 70
Shelter price indices for Canada, 1961–81

	CPI shelter components								
	Renters		*Owners*						*All items*
Year	*Rent*	*SCRI*	*Tax*	*Repr*	*Mort*	*Repl*	*SCOI*	*SCHI*	*CPI*
1961	82.4	81.6	63.1	63.4	50.4	57.5	57.4	73.1	75.0
1962	82.6	81.9	65.6	63.6	52.2	59.0	59.0	74.0	75.9
1963	82.9	82.1	67.1	65.0	54.3	61.0	60.8	74.8	77.2
1964	83.2	82.6	69.3	67.7	55.9	64.5	63.3	76.0	78.6
1965	83.7	83.2	71.8	70.7	57.8	67.6	66.0	77.3	80.5
1966	85.1	84.6	75.2	73.4	60.3	70.7	68.9	79.5	83.5
1967	87.9	87.4	78.2	77.9	63.2	75.7	72.8	82.9	86.5
1968	91.7	91.3	83.4	82.7	68.8	81.0	78.1	86.7	90.0
1969	95.3	94.9	90.7	86.0	78.6	87.2	85.1	91.2	94.1
1970	98.4	98.2	97.3	93.6	89.1	92.9	92.5	95.7	97.2
1971	100.0	100.0	100.0	100.0	100.0	100.0	100.0	100.0	100.0
1972	101.2	101.5	101.6	107.6	108.7	110.6	108.0	104.7	104.8
1973	102.6	103.2	103.5	118.3	120.9	125.1	118.8	111.4	112.7
1974	105.4	106.2	104.0	136.6	136.7	138.4	130.3	121.1	125.0
1975	111.1	112.0	111.2	146.4	156.8	146.2	143.6	133.2	138.5
1976	118.9	120.0	125.7	163.6	179.1	164.2	163.4	148.1	148.9
1977	126.3	127.6	141.1	178.8	198.8	179.3	181.2	161.9	160.8
1978	132.9	134.4	153.9	193.2	215.0	193.0	196.1	174.1	175.2
1979	138.9	140.6	161.8	214.8	228.0	200.9	208.3	186.2	191.2
1980	145.5	147.6	166.6	231.4	251.2	216.0	223.9	201.4	210.6
1981	154.8	157.4	173.0	258.4	296.2	244.7	252.5	226.4	236.9

SOURCE: Statistics Canada, *Prices and Price Indexes*, various years

hold (i.e., dwelling) . . . enters the sample, . . . data are collected during . . . six consecutive months from whatever household resides at that location. One sixth of [the sample] is replaced each month . . . A special questionnaire is provided on rent, dwelling features and services included in rent . . . Quality adjustments are made to rents, dwelling by dwelling, taking into account the estimated values of added or discontinued services; e.g., newly-provided stove, discontinuance of "free" cablevision. The ratio of average rents over two consecutive months, computed for the same dwelling sample . . . is the month-to-month rent index (Statistics Canada. 1978. *The Consumer Price Index: Revision Based on 1974 Expenditures: Concepts and Procedures*, Catalogue 62-546, 35–6).

Let us call this index the Statistics Canada rent index (SCRI). Postwar levels of SCRI (based on 1971 = 100) are shown in Table 70.

Some interesting features of SCRI are highlighted in the above quotation. To understand the methodology, consider a simpler alternative, ARI. One month, take a national random sample of renters. Calculate the average rent paid. Let this be the base rent. Repeat the sampling and

TABLE 71
Hypothetical data: number of dwellings and average rent by dwelling type and month, case A

	Number of dwellings			*Average rent*		
	May	*June*	*July*	*May*	*June*	*July*
Type A	1,000	900	800	100	101	102
Type B	1,000	1,200	1,400	200	202	204
Total	2,000	2,100	2,200	150	159	167
ARI (May = 1.0)				1.00	1.06	1.11

average rent calculation in succeeding months. Let ARI be the ratio of the average rent in a succeeding month to the base rent. Imagine a hypothetical example in which all rents increase at 1 per cent a month, and where rental dwellings are of just two kinds: low-quality (type A) and high-quality (type B). Suppose further that all A units have the same rent, as do all B units. Over a three-month (May to July) period, suppose there is a substantial addition of B units as shown in Table 71. Using the three average rents (150, 159, and 167), ARI increases by 6 per cent between May and June, and 5 per cent between June and July. ARI overstates the actual 1 per cent monthly increase because of the changing mix of A and B units. If the mix had stayed the same, ARI would have gone up by just 1 per cent a month.

How does SCRI avoid mixing quality and price change? Table 72 illustrates the mechanics, using a hypothetical breakdown of the sample by duration in the survey. As shown, suppose that 833 A dwellings in the sample in June were also there in May. Each of these had a June rent of $101 versus $100 in May, by our assumption above. Suppose there were also 833 B dwellings in a similar position and that their rents were $202 for June and $200 for May. Thus, for every dwelling in our sample, regardless of type, the ratio of June to May rent is 1.01. Hence, SCRI will be 1.01. By a similar argument, between June and July, the rent ratio will again be 1.01 for every included A or B dwelling. Hence, SCRI rises an additional 1 per cent, corresponding to the pure price change.

SCRI may substantially underestimate rent increases because it does not consider the change in rents associated with new buildings. Consider another hypothetical example in which all rental units are currently of type A, with the rents given in the previous example. Suppose a new type of rental unit (B) is constructed that is of the same quality but has a higher initial rent. Suppose that landlords gradually replace all A dwellings with B units but that rents for both increase at 1 per cent monthly. What happens to SCRI is illustrated in Table 73. SCRI rises by just 1 per cent a month. It misses the jump in rents as households relocate from A

TABLE 72
Hypothetical data: dwellings by type, month, and duration in sample, showing average rent, case B

	Type A *dwellings*			*Type* B *dwellings*			*All dwellings*		
	May	*June*	*July*	*May*	*June*	*July*	*May*	*June*	*July*
In Sample	1,000	900	800	1,000	1,200	1,400	2,000	2,100	2,200
1st Month	166	67	67	166	367	367	332	434	434
2nd Month	167	166	67	167	166	367	334	332	434
3rd Month	167	167	166	167	167	166	334	334	332
4th Month	166	167	167	166	167	167	332	334	334
5th Month	167	166	167	167	166	167	334	332	334
6th Month	167	167	166	167	167	166	334	334	332
Subsample in SCRI	-	883	733	-	833	1,033	-	1,666	1,766
Avge subsample rent									
This month		101	102		202	204			
Last month		100	101		200	202			
SCRI								1.01	1.02

to B units. To the extent that SCRI understates pure price increases, it is primarily for this reason.[4]

However, SCRI may also overstate price increases. SCRI tends to measure as a price increase changes in rent associated with quality improvements. The Rent Survey does control for some quality change; it asks about the availability of (and charges for) parking facilities, water, electricity, cablevision, major appliances, draperies, and carpeting. Where a change in these is indicated, an attempt is made to control for the impact on

TABLE 73
Hypothetical data: dwellings by type, month, and duration in sample, showing average rent, case C

	Type A *dwellings*			*Type* B *dwellings*			*All dwellings*		
	May	*June*	*July*	*May*	*June*	*July*	*May*	*June*	*July*
In Sample	600	550	500	100	200	300	700	750	800
1st month	100	50	50	100	100	100	200	150	150
2nd month	100	100	50		100	100	100	200	150
3rd month	100	100	100			100	100	100	200
4th month	100	100	100				100	100	100
5th month	100	100	100				100	100	100
6th month	100	100	100				100	100	100
Subsample in SCRI	-	500	450	-	100	200	-	600	650
Avge subsample rent									
This month		101	102		202	204			
Last month		100	101		200	202			
SCRI								1.01	1.02

rent. However, other quality changes are ignored in the Rent Survey. A landlord might, for example, improve the building insulation, redo the plumbing or wiring systems, repaint or sandblast the building, install an emergency fire escape system, or improve the building security system. Rent increases associated with such quality improvements could mistakenly be identified as price change by SCRI.

In short, SCRI may be biased either upwards or downwards on net. There is a need to be cautious in treating SCRI as a pure price index. Further, because price increments for new rental housing are not included, SCRI may have subtantially understated pure price change unless there was a substantial simultaneous improvment in dwelling quality.

Shelter Price Index for Owners

Statistics Canada produces, as part of the CPI, a price index for home owners. As with SCRI, the method of calculation changed between 1947 and 1981. The following method was employed in later years:

> In the case of owned accommodation, the 1974-based CPI series does not refer to the basket of purchased dwellings but to the stock of owner occupied dwellings in 1974. The weight attributed to this series includes the following elements of the estimated cost of owning and using the dwelling in that year:
>
> (a) the replacement cost of the amount of owner-occupied dwellings used up in 1974; such cost, equal to the annual depreciation of the owner-occupied dwellings, is estimated to be 2% of the 1974 market value of the stock (the value of land is not included);
> (b) mortgage interest paid by home owners in 1974;
> (c) property taxes paid by home owners in 1974;
> (d) cost of insurance on owner-occupied dwellings in 1974;
> (e) cost of repairs made and contracted by home owners in 1974
>
> (Statistics Canada. 1978. *The Consumer Price Index: Revision Based on 1974 Expenditures: Concepts and Procedures*, Catalogue 62-546, 24).

Let us refer to this as SCOI, or Statistics Canada's owners' (expense) index. Postwar changes in SCOI and its five components are described in Table 70. Between 1961 and 1981, SCOI quadrupled, while SCRI only doubled. The gap between the two indices increased steadily during this time. On the basis of this, it has been argued that owner-occupiers faced a worsening affordability problem in contrast to renters, who found housing becoming more affordable.

Several features of SCOI stand out. First, it is not a price index for new housing. Statistics Canada has, since 1971, published a new housing price index (NHPI).[5] "[NHPI] reflect[s] pure price change for new houses sold by major builders. Such indexes . . . reflect . . . the costs of land, labour, and materials which the builder must pay to build an identical structure in the same or comparable location . . . The indexes do not reflect . . . changing . . . quality . . . Thus costs . . . excluded from affecting price movement [include]: model substitutions, changes in the size and location of building lots, design and some construction technique changes and the provision of certain extra features such as appliances" (Statistics Canada. 1975. *Construction Price Statistics: New Housing Price Indexes: 1971 = 100*. December 1974. Catalogue 62-007, 5). In principle, this is a constant quality price index. However, building materials, dwelling design, construction techniques, and lot layouts changed in later decades. It is unclear how one could control for such quality changes entirely. Presumably some changes did not alter quality (if we recognize that different people may view the change differently), while others did. To define a constant quality price index, one needs many criteria to define when a given change in dwelling construction technique affects quality.

Values for NHPI since 1971 are roughly those shown in the column titled "Repl" in Table 70. NHPI increased more rapidly than did SCOI between 1971 and 1974 but rose less quickly thereafter. Why? SCOI measures changes in occupancy cost. It includes, but is not limited to, the selling price of a new dwelling. It more resembles the cashflow cost of home ownership. It includes mortgage interest, property taxes, home insurance, and repairs, items that are not included in NHPI.[6] It also excludes capital gains on the sale of a dwelling and the imputed rent on owner's equity. However, unlike a cashflow cost, SCOI excludes mortgage principal repayment and includes a replacement cost that is an allowance for depreciation.

Why is SCOI so complicated? Why not simply add up what a consumer spends to maintain a dwelling of fixed quality during a year, i.e. a cashflow cost? There are several problems with a cashflow approach to owner-occupancy. That owner-occupied housing is both a consumption good and a capital good is one problem. What a renter pays for housing in the private market is generally the value of that housing as a service to be consumed during a short period of time.[7] However, what an owner pays for housing reflects both the value of the service to be consumed and an expectation of capital gain or loss at some future date.

Added to this is a complexity arising because of income taxes. In Canada, there is no taxation of capital gains on the disposition of a principal residence or the imputed rent on owner equity. A renter who, in contrast, used the same money to purchase other assets did pay tax on any gains (at least after 1972). Further, as one's marginal tax rate rises, the after-tax cost of owner-occupancy declines. Before the mid-1970s, tax rates were not adjusted for inflation. Hence even households experiencing no real gain in income saw their marginal tax rate climb. Over time, this "inflationary creep" reduced the after-tax cost of home ownership. Thus, in measuring the price of owner-occupied housing, it is important to look at "after-tax user cost" (ATUC), the net cost of housing to the consumer, taking taxes into account. Like ATUC, SCOI excludes principal repayment in calculating mortgage expense; this is simply a transfer between assets, not an expense.[8] It also takes account of building depreciation.[9] However, it considers neither the imputed rents on owner equity nor the benefit of tax-free capital gains.

An alternative price index for owners has been proposed by McFadyen and Hobart (1978). Their measure of user cost, MHUC, takes the form:

$$\text{MHUC} = (\text{r} + d)P + X - G, \tag{23}$$

where P is the value of the dwelling at date t, r the "appropriate rate of return to the asset," d the rate of depreciation for housing, X property taxes plus insurance plus maintenance expenditures, and G the change in house price (i.e. capital gain or loss) over the period.[10] Their formulation differs from SCOI because it includes capital gains, the opportunity cost of owner equity, and possibly a differential between the mortgage rate and the "appropriate rate of return" on mortgage principal. Between 1961 and 1976, they find that capital gains dominated MHUC. In 1974, at the height of a housing price boom, capital gains were so large that MHUC became negative. This suggests that capital gains may be an important omission in SCOI, and hence the overall consumer price index. In other words, SCOI overestimates the increasing cost of home ownership, at least in the 1960s and 1970s, and possibly in earlier decades as well.

Similar findings have been reported in the United States. Hendershott and Shilling (1982) analyse changes in the "user cost of capital" between 1955 and 1979. User cost of capital is the annual cost to a household of a dollar invested in owner-occupied housing. They find that the expected rate of inflation of house prices rose steadily from a trough about 1963 to reach a peak of 10 per cent (per annum) in 1979. As a result, they argue that the after-tax user cost of capital declined substantially during the 1960s and 1970s. Diamond (1980, 295) draws a similar conclusion: "Ownership costs as a percent of dwelling price . . . decline[d] over the period 1970–1979. This decline is greater than the . . . rise in the cost of new dwellings. [For] the median husband-wife household considering the consumption of a marginal unit of housing services, the after tax cost . . . declined by 30 per cent from 1970 to 1979. This [reflects] a decline in the before-tax real cost of capital and a large inflation-induced increase in the tax subsidy to owner occupied housing. The . . . decline is qualitatively supported by evidence that new homes sold have become significantly larger and better appointed ... and that homeownership has become more prevalent."

With the exception of a few brief periods, notably around 1960, house prices rose steadily in Canada. Also, over the postwar period, the typical marginal tax rate increased, reflecting both rising real incomes and inflationary creep. Because of these two trends, SCOI substantially overstates the after-tax user cost of housing.

SCOI controls for quality change in part by measuring price changes for the fixed stock of dwellings in the base year expenditure survey. Further, SCOI uses a number of quality-controlled price indices to estimate changes in its components. For example, NHPI was used to estimate mortgage interest expense and (in later years) replacement expense on the assumption that the price of existing housing follows the price of new units.[11]

However, in some respects, SCOI does not control for quality change. There is no adjustment for quality change in the property tax expense component. It is assumed that changes in property taxes reflect pure price movement, not a change in service (hence quality of dwelling). In later decades, most local governments improved services. Sidewalks, paved roads, street lights, sanitary sewers, piped water, libraries, schools, transit services, and social programs (to name just a few) became almost ubiquitous. These goods and services add to the quality of a dwelling. However, SCOI treats their cost as a pure price change, hence overstating the increase in price. Also, remember that SCOI is based on the fixed stock of dwellings in the base year expenditure survey. As long as the base stock remains the same, SCOI controls for quality change. However, each time the base changes (e.g. from 1974 to 1978), so does the mix of dwellings. This is another example of the problem of changing bundles in intertemporal price index comparisons.

How good a price index is SCOI overall? In some respects, SCOI is not bad. Within the period between any pair of base surveys (i.e. reweightings), this index does control for quality change in some significant ways. However, it is unclear whether SCOI captures correctly the user cost of housing. By ignoring capital gain and the nontaxation of imputed income, SCOI overstates price increases; and the extent of this bias grew in the 1950s, 1960s, and 1970s. On net, SCOI probably overstated the actual postwar price trend for owner-occupied housing.

THE RISE IN PROSPERITY

By almost any measure, Canadians experienced unprecedented economic growth in the postwar period. Rapid expansion of employment was accompanied by much new investment. Consequently, both wage and nonwage incomes rose, and there was substantial accumulation of wealth. Whether one uses aggregate employment, output, income, wealth, or consumer expenditure as the measure, the trend was strongly upward.

Table 69 illustrates this using data on per capita personal disposable income (PCPDI). Note that these data are in current dollars, unadjusted for inflation. PCPDI rose about 12-fold, more than twice the increase in the CPI. Thus incomes grew substantially even after adjusting for inflation.

Why did incomes rise so fast? A full answer is beyond the scope of this book. However, some contributing factors can be readily identified. The accumulation of capital in plant and equipment, the development of improved management techniques, and new technologies were important. These generated new nonwage income in dividends, interest, rent, retained earnings, and noncorporate profits. That new investment, abetted by improved education and skills-training, boosted labour productivity, and hence wages. In addition, it permitted lower prices, which in turn boosted domestic and export sales, which also led to more employment. Further, much of this new investment took place in industries where labour productivity, and hence wages, were higher. Thus total wage income grew as employment declined in older low-productivity industries and rose in new high-productivity sectors. Another important factor was the increasing level of education and skill of the workforce. Being better trained, the workforce was more productive and thus earned higher incomes.

Complementing these arguments is the significance of changing dependency ratios. PCPDI is the ratio of aggregate personal disposable income of a nation to its total population. For most people, employment is, by far, the single largest source of income. Typically, the greater the proportion of population employed, the higher will be incomes and thus PCPDI. The proportion of population not in the paid workforce is the dependency ratio. To varying degrees, this dependent population is made up of three principal categories: children, the retired elderly, and nonworking wives. Over the postwar period, the size of each group changed.

With the postwar baby boom and subsequent baby bust, the number of children increased steadily until about 1970, then fell. Combined with this, the increased time spent in school reduced full-time workforce participation among young adults, although this was partly offset by a rise in part-time participation. These changes slowed the rate at which PCPDI increased, especially up to the mid-1960s. With increasing longevity, the number of people surviving past retirement age grew steadily. Earlier and more widespread retirement further boosted the number of nonworking elderly, further reducing the rate of increase of PCPDI.

Finally, there was the impact of increasing workforce participation among wives. In the early postwar period, participation rates among married women aged 25–54 were low. Only among younger (typically pre-child bearing) and older (post-child bearing and pre-retirement) wives was there much paid employment. This began to change in the mid-1950s, at about the time that marriage (and then fertility) rates fell. Many wives moved into the paid workforce, pushing up PCPDI. Through the 1960s and 1970s, the increased participation of women was an important source of growth in PCPDI.[12]

However, it is misleading to look at housing costs or affordability simply in terms of individual income. Most people do not consume housing alone. They live with their family, other relatives, friends, or partners with whom living expenses are typically shared. Therefore, the group's income helps to determine the kind of housing consumed.

For several decades, Statistics Canada has reported the average income of a "spending unit," hereinafter called AUY. A spending unit is roughly equivalent to an economic family, although it can also consist of unattached persons.[13] An index series (1971 = 100) for AUY is presented in the right-hand column of Table 69.[14] In the 1950s and 1960s, AUY increased faster than PCPDI but fell behind in the 1970s.[15]

In addition, the distribution of income among spending units shifted because of the changing mix of consumers. One important change was the rise of the working wife. As a consequence, the average income of husband-wife families rose relative to other kinds of consuming units. Another important change was the proliferation of nonfamily individuals and persons living alone. Typically, these individuals had lower incomes than other kinds of household heads. A third change was the rise of the young lone-parent family. Predominantly headed by women, such families had very low incomes.

The Working Wife

Husband-wife families, especially those in their child rearing years, made substantial gains. They did better than most other types of spending units, primarily because of rising workforce participation. To see the effect this had on AUY, consider Table 74 which shows index series for real average gross incomes of individuals, families, and spending units by sex of head.

Look first at the data for individuals. Among men with any income, average real incomes doubled between 1951 and 1981. Because of increasing workforce participation among wives and improved pension benefits for the elderly, the incomes of women rose still faster. Thus

TABLE 74
Index of average constant-dollar incomes of individuals, families, and spending units, Canada, 1951-81

Individuals or heads	Index of average constant-collar incomes (1951 = 100)*				1981 average income (1981$)
	1951	*1961*	*1971*	*1981*	
(a) Man or male head†					
Individual income‡					
All individuals	100	134	180	200	18,516
Head of spending unit					
All heads	100	132	187	208	21,209
Unattached individuals	100	131	197	263	16,239
Family income§	100	135	199	249	31,884
(b) Woman or female head					
Individual income					
All individuals	100	140	183	250	9,522
Wife of spending unit head	100	142	203	273	9,599
Head of spending unit					
All heads	100	156	215	291	11,638
Unattached individuals	100	158	239	320	11,430
Family income	100	136	157	206	18,364

SOURCES
Statistics Canada. 1969. *Incomes of Nonfarm Families and Individuals in Canada, Selected Years 1951–65*. Statistics Canada
– 1973, 1983. *Income Distributions by Size in Canada*. Statistics Canada
* The 1951 and 1961 data are based on surveys of nonfarm spending units only. The 1971 and 1981 data cover all spending units.
† A husband, where present, is defined to be the head of family. A woman is a head only when there is no spouse present.
‡ Data on individual incomes are only for persons with income. Individuals without income are excluded. All figures are deflated by the consumer price index so that 1951 = 100.
§ An economic family of two or more persons. Persons not part of an economic family are defined to be "unattached" individuals. A spending unit can be either an economic family or an unattached individual.

although women still had much lower incomes than did men in 1981, their position improved relatively.

Now, consider how income changed by type of living arrangement. The rapid growth in total family income among male-headed units reflects the increasing incidence of working wives. Between 1951 and 1981, only about two-thirds of the increase in total family income is accounted for by the rise in husband's income. Much of the rest was attributable to wives. In contrast, total family income increased slowly among female-headed (typically lone-parent) spending units.

As important as the amount of income generated was the contribution

TABLE 75
Average constant-dollar incomes of male-headed economic families by age of head, Canada, 1971 and 1981

Age Group	*Avge family* income ($)* 1971	1981	*1981 CPI-adjusted index† (1971 = 100)*	*Coeff of variation‡* 1971	1981
Under 25	8,248	23,870	122	0.6	0.6
25-34	10,435	30,190	122	0.7	0.7
35-44	11,895	35,581	126	1.0	0.7
45-54	12,361	38,715	132	0.8	0.8
55-64	11,183	32,753	124	1.5	0.9
65-69	7,901	23,117	124	1.0	1.0
70 or older	5,808	18,162	132	1.3	1.0

SOURCE: Statistics Canada. 1973, 1983. *Income Distributions by Size in Canada*. Statistics Canada

* An economic family is a group of two or more coresident individuals related by blood, marriage, or adoption. These data cover all economic family spending units in Canada.

† Ratio of 1981 to 1971 income divided by the (1971 = 1.0) CPI for 1981, which stood at 2.369

‡ Ratio of standard deviation to average family income in the age group

of working wives to the distribution of wealth. Typically, the working wife was a second source of employment income. Most of the increase in working wives was in the middle (25-44) age groups. These, however, are for the most part also the peak earnings ages of their spouses. In other words, what were already well-off households were the principal beneficiaries of increased participation by working wives. In contrast, low-income units, such as the elderly, lone parents, and people living alone did not benefit directly from this increased participation.

Did this lead to increasing income inequality among families? J.P. Smith (1979, S166) found little change in income inequality between 1960 and 1970 among U.S. husband-wife families in the age groups where female participation rates rose the most. Smith suggests that wives' earnings offset those of their husbands. Because increased participation came mainly from wives whose husbands had low incomes, it did not add to the level of inequality.[16] However, Smith did not compare husband-wife families to other kinds of households, and so we cannot say how the level of income inequality changed when all households are considered.

There is some indirect evidence for Canada. In Table 75 are shown the age-disaggregated average incomes of male-headed economic families in 1971 and 1981. Although not exactly the same, male-headed economic families are preponderantly also husband-wife nuclear families, and in what follows they are assumed to be the same. In Table 75, the changes in average income are remarkably similar across age groups. Over the 1970s, real average family incomes rose by 22 per cent to 32 per cent across all age groups. However, because female participation increased most rapidly in the 25–54 age groups, one might have expected larger increases in the corresponding age groups of husband. Apparently, the income gains among the middle-aged were not much higher than those at other ages. Increased female participation did not lead to more pronounced inequality across age groups. The right-hand columns show the variation in family income within each age group in 1971 and 1981. Over the 1970s, variation declined substantially among families headed by men aged 35–44, 55–64, or 70 or older. In the case of the first age group, this might reflect increasing workforce participation among wives, as it is consistent with the argument that working wives tended to reduce the dispersion of family incomes. However, among other age groups, there was little change in the dispersion of incomes, supporting the argument that increased female participation did not cause greater inequality.

The Unattached Individual

Equally evident in Table 74 is a sharp rise in income among unattached individuals. Their incomes grew faster than those of family heads, and even total family incomes. Although unattached individuals were still typically far worse off than men heading families, they were better off than others who were members but not heads of families. Between 1951 and 1981, this increasing affluence permeated all age groups. See Table 76. Real incomes trebled. Only among 55–64-year-olds was the appreciation somewhat less, although appearance is deceiving in this case. Over the 1970s, the number of 55–64-year-old unattached women increased by about one-third; the corresponding number of men stayed about the same. Thus, the sluggish growth of income in this age group reflects the greater incidence of women with correspondingly lower incomes. In fact, considering each sex separately, incomes grew substantially.

Chapter four describes the rapid formation of-one person households. By definition, everyone living alone is an unattached individual. The latter also includes anyone living with others who are unrelated by blood, marriage, or adoption. Table 77 contrasts estimates of the number of unattached individuals in Canada with census counts of persons living alone.[17] By 1981, two-thirds of all unattached individuals were living alone. In chapter six, choice of living arrangement is shown to be especially sensitive to income among nonfamily individuals. Indeed, in cases such as home leaving among young adults, the decision to become an unattached individual is presumably contingent upon being self-supporting.

TABLE 76
Index of average real incomes of unattached individuals by age, Canada, 1951–81

Age group	*CPI-adjusted average individual income (1951 = 100)**				*1981 average income (1981$)*
	1951	*1961*	*1971*	*1981*	
Under 25	100	161	200	290	11,500
25–34	100	151	235	280	17,492
35–44	100	149	242	311	19,317
45–54	100	145	233	290	16,436
55–64	100	135	223	257	12,591
65 or older	100	160	198	287	9,500

SOURCES
Statistics Canada. 1969. *Incomes of Nonfarm Families and Individuals in Canada, Selected Years 1951–65.* Statistics Canada
– 1973, 1983. *Income Distributions by Size in Canada.* Statistics Canada
* An economic family is a group of two or more coresident individuals related by blood, marriage, or adoption. An unattached individual is anyone who is not a member of an economic family. Figures include only individuals with any income; individuals without income are excluded. Index values are deflated by the consumer price index so that 1951 = 100. The 1951 and 1961 data cover nonfarm spending units only; the 1971 and 1981 data cover all spending units in Canada.

The increases in income concurrent with the rise of the one-person household are somewhat surprising. Much of the new formation of one-person households was among the less affluent – seniors (especially widows) and young singles. Further, among the elderly, part of this increase came about because of publicly assisted housing, where low income was a precondition to admission. Given this demand pressure and the moderate increase in rents suggested earlier in this chapter, one might have expected sluggish growth in income. Instead, incomes rose relatively quickly. Why?

In part, it reflects the changing character of the one-person household. As reported in the 1951 Census (vol 10, 368): "The highest percentages of one person households were found in rural non-farm areas, and . . . were much more common . . . west of the Great Lakes . . . It is probable that a fair percentage of them consisted of hunters, trappers, west coast fishermen, fire rangers, guides, and persons in similar occupations." Thus, in 1951, many people living alone were in low-wage occupations wherein living arrangement was dictated by one's job and likely subsidized by the employer. Further, even where not subsidized, the housing was likely of poor quality, and hence inexpensive. During the following three decades, the character of the one-person household changed. It became primarily an urban phenomenon, a living arrangement chosen

TABLE 77
Unattached individuals and persons living alone (000s), Canada, 1951-81

	1951	*1961*	*1971*	*1981*
Unattached individuals	582	754	1,729	2,544
One-person households	252	424	812	1,681

SOURCE: Computed from published census reports, various years

by an individual independent of his or her occupation. Living alone, without subsidy, was not inexpensive. Further, stricter implementation of building codes and zoning bylaws in urban areas meant that such housing was more expensive than that typically occupied by someone living alone in 1951. As a consequence, the minimum income required to live alone increased substantially.

A second factor was the introduction and extension of universal pension plans. Elderly widows, the most prevalent kind of one-person household, experienced great gains in income in the 1960s and 1970s. Important too were the inflation-indexed provisions of public programs, including Old Age Security, the Guaranteed Income Supplement, the Survivor's Pension, and the Canada/Quebec Pension Plan.

Another consideration, in the 1960s and 1970s, was the effect of the marriage bust. During the marriage boom of the 1950s, relatively few young adults ever lived alone. Even the most upwardly mobile tended to marry, and marry early. Those who did not marry included many who were relatively less concerned with acquiring income or a career. However, with the marriage bust of the 1960s, things changed. It was no longer as common for the upwardly mobile to see marriage as part of a life path. Instead, the ranks of young adults began to include more singles for whom incomes and career advancement were important. In other words, the income of a typical person living alone rose in the 1960s and 1970s because that typical person was different from his or her predecessor.

The Lone Parent

The incomes of female lone-parent families increased relatively slowly. Although we cannot see this directly in Table 74, we can use as a proxy the incomes of female-headed families, as most of these units are in fact lone-parent families. Slightly more than doubling in three decades, their incomes lagged far behind those of male-headed families. Closer examination shows that the slow rate of increase was restricted to the 1960s. In the 1950s and 1970s, female-headed family units experienced gains not

TABLE 78
Average incomes of female-headed economic families by age of head, Canada, 1971 and 1981

	Avge family income ($)		*1981* CPI-*adjusted index**
Age group	*1971*	*1981*	*(1971 = 100)*
Under 35	3,668	11,542	133
35-44	5,564	19,237	146
45-54	6,660	22,459	142
55-64	6,264	22,180	149
65 or older	7,445	22,368	127

SOURCE: Statistics Canada. 1973, 1983. *Income Distributions by Size in Canada*. Statistics Canada
* Ratio of 1981 to 1971 income divided by the 1971-based CPI for 1981, which stood at 2.369

unlike those of male-headed units. What caused the decline in relative position in the 1960s?

One possible explanation is the rise of the working wife that began in the 1960s. This caused the incomes of husband-wife families to increase faster than those of lone parents because labour force participation did not rise correspondingly for the latter. It also explains why the incomes of female-headed families increased less quickly than for wives, although the latter still tended to be considerably smaller than the former. This explanation does not fit the 1970s very well, though. If anything, labour force participation among wives increased more rapidly than during the 1960s. However, female-headed family units experienced larger gains in income. Further, these large income gains were experienced at all ages. See Table 78. In other words, the gains were not simply the result of a shift in the age distribution such as aging of the baby boomers.

INCOMES, PRICES, AND HOUSING DEMAND: AGGREGATE PERSPECTIVE

To what extent did postwar housing consumption reflect prosperity? To what extent did it also reflect the changing price of housing relative to other goods? In this section, we consider aggregate estimates in which changes in demographic composition are ignored or treated crudely. The ensuing section considers the role that demographic changes may have played in conjunction with income and price.

Three aggregate housing market models are considered. The first, by Muth (1960), models U.S. housing consumption in the period between the two world wars. Although not concerned with postwar change, this seminal paper offers a unique perspective and some interesting comparisons with the following two. The second, by Hickman (1974), models housing consumption in the United States between 1922 and 1970. The temporal overlap with Muth provides insights. Also, although it is not concerned with Canada, there are enough similarities in the U.S. postwar experience to make the study of interest here. The third, by L.B. Smith (1974), models housing consumption in Canada between 1950 and 1970. These three studies illustrate the potential of an aggregate approach.

Muth's Model

Muth (1960) describes a complex analysis of U.S. housing demand in the interwar period. In this review, just two of his estimated equations are considered. One takes the following form:

$$h'_f = f(p, y, r, h), \tag{24}$$

where h'_f is the gross rate of nonfarm residential construction discounted by the Boeckh index and divided by nonfarm population,
p is the Boeckh index of residential construction costs (brick),
y_p is Friedman's per capita expected-income series,
r is Durand's basic yield of ten-year corporate bonds, and
h is the beginning-of-year per capita housing stock.

This models the annual real per capita volume of residential construction, including both new dwellings and additions or alterations to existing units. Muth divides by a price index to get an aggregate price-discounted expenditure on housing construction. Then he divides by population size to get a per capita figure. Muth sees (24) as a demand curve for residential construction with three independent variables to proxy shelter prices (for both construction and financing) and income. The last variable (h) allows for incomplete adjustment in the housing market.

Muth concludes that housing demand is quite sensitive to both price and income change. From multiple regression estimates, he finds that the long-run (i.e. equilibrium) effect of a 1 per cent increase in price is a 0.9 per cent reduction in the per capita real dollar value of the housing stock. A 1 per cent rise in income, however, implies a 0.9 per cent rise in the housing stock.

Muth also examines the price and income elasticity of dwelling quality. The total dollar value of the housing stock is the product of the number of dwellings and the average dollar value per dwelling. When deflated by a price index, the latter corresponds to a price-discounted quality measure of the form $h = e/p$. Muth estimates a second equation (equation [27], 67)

$$H_n = f(p, Y_p, r, S), \tag{25}$$

where H_n is the average current dollar value of a new dwelling unit discounted by the Boeckh index,
Y_p is average expected income per household, and
S is the average size of household.

The dependent variable is the average quality (i.e. price-deflated expenditure) of a new dwelling. It differs from the previous dependent variable in that it is restricted to new dwellings and is the average expenditure per dwelling constructed, rather than per capita. Muth finds that the income elasticity of new dwelling quality is about 1.9. The price elasticity is estimated to be about −1.2. He concludes that new dwelling quality is very sensitive to price and income.

Average household size has a positive coefficent in Muth's quality equation. Thus, as household size declined, so too did average dwelling quality, other things being equal. However, household size is not statistically significant. In other words, Muth's findings are not inconsistent with the view that variations in household size did not have a large effect on price-discounted housing expenditure.

Hickman's Model

Hickman's model consists of two parts. The first models household formation and the aggregate stock of dwellings. It takes the following form:

$$\mathrm{HU}_t = f(p/z_t, \mathrm{HU}_{t-1}), \tag{26}$$

$$\mathrm{HH}_t = g(p/z_t, \mathrm{HHS}_t, y/\mathrm{HHS}_t), \tag{27}$$

where HU is the number of nonfarm dwelling units,
p is the price of housing (implicit price deflator for actual or imputed rent of non-farm dwellings),
z is the price of all goods in general (implicit price deflator for personal consumption expenditure),
HH is the number of nonfarm households,
HHS is the number of standardized households (found by applying U.S. age-specific headship rates for 1940 to the estimated

national population by age group in each year),
y is real personal disposable income, and
t is the date (year).

The first equation ties the supply of dwellings in year t to the relative price of housing and the supply carried over from the preceding year. The second relates the demand for housing units (here equal to the number of households) negatively to the relative price of housing and positively to the number of "standardized" households and income per standardized household. Hickman also introduces a market equilibrium equation,

$$HU_t = HH_t, \tag{28}$$

and a mechanism whereby HU and HH can be in disequilibrium while at the same time slowly adjusting over time toward an equilibrium defined by (26), (27), and (28).

These equations model the numbers of dwellings and households. There is a close correspondence between the two. Since each occupied dwelling contains one household, they diverge only in that HU includes unoccupied dwellings. Hickman relates household formation to the age structure of the population using the variable HHS. HHS indicates the "normal" number of households formed, based on 1940 headship rates. In (27), actual household formation can be larger than this when the relative price of housing (p/z) is low or income (Y/HHS) is high, or smaller in the opposite situation.

The second part of Hickman's model predicts the number (HSPU) and average quality (HCCA) of new private housing starts. Hickman relates HSPU to profits expected by builders, higher profitability implying a larger supply. Explanatory variables include the ratio of expected selling price to construction costs, the vacancy rate, household disposable income, and credit availability. HCCA is the real average value of a private housing start. This is a quantity measure of the amount of housing of the form $h = e/p$, similar to Muth's quality measure.[18] In Hickman's view, HCCA is determined by builder profitability. He relates HCCA to real personal disposable income per household, an interest rate variable, average household size, and the typical constant dollar value of existing housing.

Hickman estimated his model using annual data for the United States between 1922 and 1970. Results in the first part of the model indicate that rent levels had almost no effect on household formation in (27). A once-only rise in p/z of 1 per cent induced, over the long run, a decline in total households formed of only 0.01 per cent. However, the effect of income was more substantial. A once-only rise in y of 1 per cent induced, over the long run, an increase in total household formation of 0.18 per cent.

There are two striking features in these estimates. First, they are much closer to zero than Muth's price and income elasticities for total housing stock change of −0.9 and +0.9 respectively. However, remember that Muth's estimates are for price-deflated expenditure, not the physical number of dwelling units. To the extent that people use their rising incomes to purchase higher-quality dwellings rather than more dwellings, these two studies need not be inconsistent.

Second, Hickman's estimated income elasticity is still large. Between 1946 and 1981, Canadian real personal disposable income rose by about 150 per cent. Applied to Canada, Hickman's work indicates that equilibrium household formation rose by 27 per cent. Household formation in Canada actually increased by about 180 per cent over the same period, but at the same time population grew by about 100 per cent. Thus Hickman's finding suggests that income growth might account for about one-third of the household formation not directly associated with population growth.

How about housing consumption? In the second part of the model, Hickman finds that a 1 per cent rise in real personal disposable income per household induced a 0.6 per cent rise in HCCA (average real value of new private housing starts). Offsetting this was the effect of declining household size. Hickman estimates that a 1 per cent decline in average household size induced a 1.7 per cent fall in HCCA. In Canada, real personal disposable income per household rose by about 70 per cent between 1946 and 1981. At the same time, average household size declined by about 30 per cent. Other things being equal, Hickman's results suggest that the increase in income should have caused HCCA to rise by 42 per cent, and the decline in household size to fall by 51 per cent, the net effect on HCCA being quite small.[19]

This finding differs from Muth's. Muth found a higher income elasticity; he also found household size to be an insignificant determinant. Why the differences? First, Hickman did not include a price variable in this equation. Because of this, the coefficient of income is pushed down; in effect, it represents the net effect of postwar rises in income and housing prices. Second, there is likely a substantial multicollinearity among these three variables. Over time, incomes went up, prices went up, and household size declined, making it difficult to separate statistically their effects on dwelling quality.

Because Hickman did not include the price of housing as a determinant of HCCA, its effect (relative to other prices or to incomes) on the quality of housing cannot be investigated. Further, although p is included in the equation for new private housing starts, its coefficient is positive, not negative. This is consistent with Hickman's view of builders responding to the profitability of new construction; the higher the price of housing, the more they construct. The only other place that housing price enters the model is in the household formation equation, and there its effect is negligible. Thus the model does not directly consider the effect on consumer demand of rising house prices.

Smith's Model

L.B. Smith (1974, chapter 4) presents a model of the Canadian housing market based on quarterly data for the early 1950s to late 1960s.[20] In its simplest form, the model contains two equations, the dependent variables being the price of housing and the volume of housing starts.

One equation (p. 49) predicts the number of housing starts:

$$\text{HS} = f(\text{PH}/\text{CLC}, \text{RM}, \text{RM-RB}, \text{CMHC}, \text{WW}), \tag{29}$$

where HS is the number of new housing starts,
PH is the arithmetic average of the average selling price of a new NHA-insured single dwelling and the average price of dwellings sold under local Multiple Listing Services,
CLC is an index of average land and construction costs per square foot of floor area of new NHA-insured single detached dwellings,
RM is the average of conventional and NHA-insured mortgage rates and the McLeod, Young, Weir 40-bond average,
CMHC is the volume of CMHC direct mortgage approvals, and
WW is a dummy to capture the effect of the Winter Works program, 1963–5.

Note in this case that the housing price variable, PH, does not control for quality change. It is not a price index like p in Hickman's work. Changes in PH reflect both price and quality variations. This makes it difficult to compare Smith's price elasticities with those found in the other studies.

As with Hickman, Smith looks at housing starts from a builder's viewpoint. The first independent variable indicates the profitability of new construction. The mortgage interest rate can also be construed as affecting profitability. The third and fourth variables are measures of the availability of private and public mortgage funding for new residential construction. In these respects, the housing starts equations of Hickman and Smith are quite similar.

The other equation (p. 47) predicts the price of housing (PH):

$$\text{PH} = f(\text{YD}/\text{FAM}, \text{SH}/\text{FAM}, \text{PGNE}), \tag{30}$$

where YD is permanent real disposable income, calculcated as a weighted average of real personal disposable income over the preceding eight quarters (two years),
FAM is the number of census families,
SH is the stock of dwellings, and
PGNE is the implicit private gross national expenditure deflator.

Smith argues that PH is negatively related to the second independent variable and positively related to the other two.[21]

Just how sensitive is housing demand to income and price in Smith's model? In the model, consumer income affects the price of housing (PH), which in turn affects the number of housing starts, and finally shows up as a change in the housing stock. Smith (p. 30) suggests that a 1 per cent rise in income leads to a 0.3 per cent increase in the number of dwellings. This finding is not far off Hickman's 0.18 per cent figure for the United States. It supports the argument that growing prosperity was an important determinant of the rise in headship rates in Canada.

INCOMES, PRICES, AND HOUSING DEMAND: DISAGGREGATE PERSPECTIVE

Aggregate approaches, as typified by Muth, Hickman, and Smith, are useful in analysing housing demand because they yield simple indicators of the importance of income and price. The three studies suggest that the income elasticity of aggregate demand for dwellings is about 0.2 to 0.3. Muth suggests that the income elasticity of demand for dwelling quality is much higher. Hickman estimates the price elasticity of demand for dwelling units to be very small. Muth finds that the typical quality of new housing constructed is quite sensitive to price. However, multicollinearity problems make it difficult to separate the effects on postwar housing demand of rising income, rising housing prices, and demographic changes such as the decline in household size. Is there any other way of separating these effects?

This section considers income elasticity estimates for different kinds of consumers from cross-sectional data. These are calculated by looking at differences in housing consumption between consumers at different incomes. The implicit assumption is that if low-income consumers become better off, their housing consumption would become like that of their more affluent peers. This assumption has never been adequately tested. As a result, caution should be exercised in interpreting cross-sectional elasticities. An observed difference in housing consumption between income groups might really be attributable to some other unobserved difference between them.

Nonetheless, cross-sectional data are useful. With them, we can disaggregate households by demographic characteristics in a way not generally possible in an aggregate approach. There is ample evidence (e.g. David [1962], Doling [1976]) that housing consumption does vary with the composition of the household, making it valuable to know the contribution of changing household structures to postwar shifts in consumption. Also, we can usually obtain more detailed information about housing consumption from cross-sectional data. Hence, we can look at the income and price elasticities of different kinds of housing demands. In what follows, we consider two demand measures: tenure and discounted housing expenditure.

Tenure

David (1962) is a seminal analysis in this area. Using the 1955 U.S. Survey of Consumer Finances, David describes how tenure was related to disposable income and spending unit size. The data in Table 79 are taken from his study (p. 58).

This table shows the sensitivity of income to tenure. Among lower-income households, the incidence of ownership was low; but it rose quickly with income. At the same time, though, note the substantial variation in tenure by size of spending unit. For example, of those earning $3,000 to $3,999 (in 1954), 21 per cent of one-person units were owners, compared with 54 per cent of two-person units. However, only among the more affluent did the incidence of owning continue to rise for households with more than two persons. Among poorer households, the incidence of ownership declined above three persons. Thus, size of spending unit, in conjunction with income, was important in determining tenure.

Why is spending unit size important? One argument is that spending unit size constrains the ability to be an owner. Consider a small and a large household at the same income. Other things being equal, the large household will spend more on "basic" consumption of food, clothing, shelter, and other necessities, because of its size. A modest income severely constrains the amount that can be saved toward a downpayment. Small households do not face the same financial pressure and can more rapidly accumulate the savings necessary to make a house purchase affordable.

A second argument is that unit size determines the housing desired, e.g. in terms of floor area or outdoor space. It is commonly argued that larger dwellings and those with private outdoor space are typically owner-

TABLE 79
Owner occupiers as a percentage of all spending units by number of persons in spending unit, United States, 1955

Disposable income of spending unit	*Number of persons in spending unit*				
	1	*2*	*3*	*4*	*5 or more*
Under $2,000	32	55	39	*	31
$2,000-2,999	17	57	41	35	27
$3,000-3,999	21	54	44	43	55
$4,000-4,999	27	57	60	56	67
$5,000-7,499	*	61	73	69	79
$7,500 or more	*	72	80	88	84

SOURCE: David (1962, p. 58)
* Sample size too small for reliable estimation

occupied. Rental dwellings, such as apartments, are often smaller or without access to a private yard. Consumers who want more space will thus be more likely to own than rent. The fly in this ointment is the presumed difficulty of creating rental units that are large or have yards or small owner-occupied units. There are few if any restrictions, though, on the ability of a landlord to rent property. In a competitive market, landlords respond to a lack of larger rental units, or dwellings with yards, by either constructing new units or converting existing dwellings from owner-occupancy to rental. The reverse argument also holds. If owners want smaller dwellings, builders can respond with more of these.[22] The space needs of different spending units are important in determining tenure only insofar as there are restrictions on the supply of particular tenure-quality combinations.

Another argument is that smaller units - especially persons living alone or couples where both partners work - do not have the time to spend on house maintenance duties. Still other small units, notably the elderly, may simply not be physically capable of such duties. An owner can purchase some of these services (e.g. grass cutting, snow removal, or house painting), obviating the need to do them himself. However, being an owner typically means at least having to worry about, and administer, such activities. Renters, however, largely need not be concerned with such things. In the 1950s and 1960s, this may have been important in the tenure decisions of small households. However, the emergence of the condominium entity in Canada in the late 1960s undermines this argument. In a condominium, some of the duties of ownership are put in the hands of a board of directors. The individual owner does not have to become involved, breaking the link between tenure and the duties of ownership.

A fourth argument concerns security of tenure. If forced to move, larger households typically have higher relocation costs. Partly, this is because they tend to have more household furnishings than a small household. Partly, it is also because large households usually contain children while small households do not. If forced to move, children may have to be relocated to new schools and seek new friends. The psychic, if not monetary, costs of such relocation can be large. It is therefore rational for a larger household to seek housing that minimizes the likelihood of a forced move. This is less important for small households, for which relocation costs are lower. For a larger household, ownership is preferable to renting on this ground. Owners are relatively secure from eviction. Over much of the postwar period, however, renters had only limited security of tenure. If a landlord wished to evict, there were typically few legal grounds on which the household could resist. It was only in the 1970s that most Canadian provinces enacted security of tenure legislation giving tenants some protection against eviction. In addition, the children in larger households themselves could be the cause of an eviction. To the extent that children create noise, or engage in rowdiness or petty vandalism, they raise the risk of eviction for the household. This provides an additional incentive for large households to become owners.

A fifth argument is that unit size is actually a surrogate for other variables. Note, for example, that 32 per cent of one-person households in the lowest income group are owners. However, in the next income group ($2,000-2,999), only 17 per cent are owners. Why isn't the incidence of owners greater in the higher-income group? Does this mean that persons living alone are less likely to be owners as their incomes go up? No. I suspect that the difference reflects age differentials between the income groups. The lowest income category includes many elderly. For the most part, they purchased the dwelling when they were younger and had higher incomes. Now, being retired, their incomes are low; however, they remain owners because the mortgage is paid off and the out-of-pocket costs of ownership are low. In the same way, one might expect substantial age variations in the other size categories. Typically, larger households are those with children; they are primarily units whose head is middle-aged. The two-person spending units, in contrast, are primarily couples, either young (pre-children) or older (post-children).

In a subsequent study of tenure in San Francisco, Straszheim (1975, 106) confirms much of the above. He finds that tenure is most sensitive to income among young singles and young marrieds with small families. Among larger and older families, the incidence of home ownership is everywhere high and does not vary substantially with income. Similar results are reported by Struyk (1976, 66) based on work with the 1970 Census for Pittsburgh. Struyk found that among husband-wife families, tenure is most sensitive to income among young and low-income families. Among older or more affluent families, the incidence of ownership is high everywhere and the income elasticity of demand correspondingly low. In related work, J.P. Smith and Ward (1980) find that the level of savings - hence the ability to afford a downpayment - may actually fall among young families, principally because of a child-induced withdrawal of wives from the labour force.

Steele (1979, chapter 6) is one of the few Canadian studies of the elasticity of demand for owner-occupancy. Her work is based on the 1971 Census public use (household) sample, the same sample used in chapters five and six. The sample contains information about each household's dwelling on 1 June 1971, tenure (owning v. renting), income (for 1970), and demographic characteristics.

Steele fits a logit model between tenure and income. As in chapter six, the logit model takes the form:

$$p = e^z/(1+e^z), \text{ where } z = b_0 + b_1 y, \tag{31}$$

where p is the unobserved probability of being an owner and y is household income. The parameters (b_0 and b_1) are chosen to make p near 1.0 in the case of an owner and near 0.0 for a renter. The income elasticity (a) of ownership is then given by:

$$a = y(1-p)b_1. \tag{32}$$

Above, (31) is a demand curve. It relates demand (tenure) to income. It does not, however, consider the effect of price on demand. In ignoring price, Steele assumes that the price differential between ownership and renting is the same across Canada. This need not be true. For example, because of capital gains, the user cost of home ownership might be lower in a local market experiencing rapid inflation. Steele ignores this problem because of the lack of house price data in the Census. However, the omitted price variable can affect the estimated relationships. For example, if high-income areas are also those with the most rapid inflation, omitting a price variable in (31) causes the estimated income elasticity to be overestimated.

In part, Steele controls for the missing price variable by disaggregation. If the households in a given subsample face the same relative prices for owner-occupancy versus renting, the estimated income elasticities should not be biased upward or downward. Steele splits the total sample of households for Canada by area of residence and age of head, estimates a separate logit model for each, then calculates the elasticity of demand at a household income of $9,391 (the average for Canada in the 1971 Census). The resulting elasticities are reported (p. 129) in Table 80. These income elasticities of ownership are highest among the young in

TABLE 80
Estimated income elasticities of home ownership, Canada, 1971

	Age of head						
Area of residence	*15–24*	*25–29*	*30–34*	*35–44*	*45–54*	*55–64*	*65+*
Urban areas							
30,000 persons or more	0.8	0.6	0.6	0.5	0.4	0.3	0.2
Small urban	0.4	0.5	0.3	0.2	0.2	0.1	0.1
Rural areas							
Nonfarm dwellings	0.1	0.2	0.1	0.1	0.0	0.0	0.1
Farm dwellings	0.0	0.0	0.0	0.0	0.0	0.0	0.0

SOURCE: Steele (1979, p. 129)

large urban areas. They decline with age and in less-populated areas.

In general, tenure in large cities is more sensitive to income than it is elsewhere. The relatively low cost of owner-occupied housing in rural areas and small towns makes ownership almost ubiquitous there. The absence, or weak enforcement, of strict building codes or zoning bylaws permit the construction of low-cost owner-occupied housing, even if of poor quality. Such inexpensive owner-occupied housing makes it difficult for a producer of rental housing to compete. In addition, there is a higher element of risk of vacancy for the landlord in a less-populated area; hence the price charged must be higher to compensate.

Why are younger households more likely to rent? One reason is that they typically do not have much accumulated wealth. Their ability to afford ownership is sensitive to, and largely dependent on, current income, i.e. on the ability to afford large mortgage payments. Older households have more accumulated wealth and thus less need for mortgage financing; their tenure is not as sensitive to current income. A second reason is the higher rate of residential mobility among young adults. A household that expects to move will take into account the expected costs of moving. For owners, this includes the cost of selling the home. A renter, in contrast, does not have to pay property transfer taxes, or substantial legal fees, when moving. Thus a household that expects to move may find it less expensive to be a renter. A third reason is that much of the tenure pattern by age of head also reflects variations in size. Young and elderly households tend to be small. Middle-aged households tend to be larger, thus more likely to own, as argued above. However, even when size is controlled, there are variations in income sensitivity among different age groups.

In Miron (1982, 67), I took two samples of two-person households, one from the same 1971 Census public use sample employed by Steele, and the other from Statistics Canada's national Household Income, Facilities

TABLE 81
Estimated percentage owner-occupiers for two-person households by type, age of head, income, and area of residence: Canada, 1971 and 1976

	30,000 persons or more				*Under 30,000/rural*			
	1971		*1976*		*1971*		*1976*	
Age of head	*X = 5*	*X = 15*	*X = 5*	*X = 15*	*X = 5*	*X = 15*	*X = 5*	*X = 15*
(a) Husband-wife								
Under 25	5	16	6	20	30	32	24	46
25-34	11	26	6	20	41	46	53	66
35-44	33	47	42	59	73	76	80	82
45-54	55	66	55	64	83	84	84	85
55-64	64	67	67	75	87	87	86	93
65 or older	63	64	66	68	87	89	89	90
(b) Parent-child								
Under 25	10	36	-†	-	-	-	-	-
25-34	12	6	13	11	-	-	-	-
35-44	18	43	9	26	-	-	-	-
45-54	30	40	30	49	62	65	60	95
55-64	46	53	46	49	80	89	81	74
65 or older	58	64	55	58	83	88	92	92
(c) Other related pairs								
Under 25	3	7	-	-	-	-	-	-
25-34	3	17	-	-	-	-	-	-
35-44	30	35	-	-	84	89	-	-
45-54	32	46	-	-	89	90	-	-
55-64	44	53	-	-	83	89	-	-
65 or older	56	56	50	55	86	98	83	94
(d) Unrelated pairs								
Under 25	1	2	4	2	13	9	1	4
25-34	2	3	5	10	-	-	-	-
35-44	16	21	-	-	-	-	-	-
45-54	16	22	-	-	32	75	-	-
55-64	50	44	-	-	73	70	-	-
65 or older	53	56	-	-	91	95	-	-

SOURCE: Computed from the 1971 Census public use sample (household file) and the 1976 Household Income, Facilities, and Equipment micro data file (calculations by J.R. Miron)
* Household income for the preceding year in thousands of current dollars
† Model not estimated because of small sample size

and Equipment (HIFE) micro data file for 1976. Each sample was subdivided by area of residence, type of household, and age of head. A logit model was estimated relating tenure to income for each group, similar to the approach of Steele. Table 81 shows the probabilities of being an owner at incomes of $5,000 and $15,000 as predicted from these models.

Among husband-wife households - panel (a) - tenure shows the most sensitivity to income among the younger age groups. Among older husband-wife pairs, the propensity to own was less responsive to income. The effect of income was still more muted or mixed among other kinds of two-person households. Among parent-child and unrelated pairs households, there are even instances of a decline in ownership at higher incomes. Further, the large differences in income sensitivity between households in large urban areas and elsewhere is noteworthy.

I also took a sample of Canadian husband-wife families with children living alone from the 1980 HIFE micro data file. These were subdivided by size of household, area of residence, and age of head. Shown in Table 82 are the estimated probabilities of being an owner at various incomes, again based on a logit model fitted to each group. In general, tenure among young families of all sizes was quite responsive to income. The tenure patterns of middle-aged and older families show smaller income elasticities, though still substantial in larger urban areas. Indeed, there is more variation in ownership among these households than there was among the two-person households in Table 81. This is consistent with David's finding that ownership did not increase steadily with household size. Particularly at lower incomes, large households were less frequently owners compared to small households; at the same time, their tenure was more sensitive to income.

These results suggest that rising prosperity should have led to an increasing incidence of owner-occupancy. What actually happened was more complex. Although stagnant for Canada as a whole, the percentage of urban households that were owners did rise between 1951 and 1961, from 56 per cent to 59 per cent. Then, in the 1960s, the incidence of ownership fell abruptly. It was only during the 1970s that ownership began to re-emerge. Why didn't rising prosperity lead to sustantially more ownership?

One answer is based on the observation that tenure is responsive to income principally among young or large families. With the baby boom, family sizes increased. However, larger families are less likely to own, holding income constant. Without a concurrent growth in income, the baby boom itself might have led to a decline in ownership. However, the higher income elasticity among larger families helped compensate for this. Rising prosperity offset the decline in ownership that might otherwise have been expected with increasing household size. In contrast, average family size shrank during the baby bust of the 1970s, contributing to a rise in ownership. At the same time, the baby boom generation began to swell the number of young adult households. This should have reduced the incidence of ownership, but the effect again was blunted by rising prosperity.

TABLE 82
Estimated percentage owner-occupiers among husband-wife families living alone by size, age of head, income, and area of residence, Canada, 1980

Area of residence, age of head	*1980 actual owner*	*Owner's estimated household income ($)*			*1971*	
		10,000	*25,000*	*40,000*	*Actual owner*	*Model estimate**
(a) Three persons						
30,000 or more						
15-34	56	26	61	87	28	46
35-54	75	59	73	84	65	74
55 or older	80	71	80	86	72	80
Under 30,000/rural						
15-34	70	57	75	87	48	65
35-54	86	82	86	89	80	85
55 or older	94	91	96	99	89	94
(b) Four persons						
30,000 or more						
15-34	61	38	74	93	47	64
35-54	85	66	83	92	74	84
55 or older	83	61	80	91	72	85
Under 30,000/rural						
15-34	84	79	86	91	58	83
35-54	91	88	91	94	82	90
55 or older	95	94	95	96	86	95
(c) Five or more						
30,000 or more						
15-34	76	58	79	91	55	75
35-54	85	64	83	93	74	86
55 or older	87	61	79	90	74	86
Under 30,000/rural						
15-34	82	81	83	84	69	82
35-54	91	87	91	94	84	90
55 or older	95	94	95	95	90	95

SOURCE: Computed from the 1971 Census public use sample (household file) and the 1980 Household Income, Facilities, and Equipment micro data file (calculation by J.R. Miron)
*Calculated from the 1980 estimated logit model using the cohort's mean household income as reported in the 1971 Census public use sample and allowing for inflation (CPI = 100 in 1971 and 212.6 in 1980)

This explanation assumes that rising prosperity was widely diffused. However, in explaining the baby boom, Easterlin (1980) indicates that the incomes of young men rose relative to those of their fathers in the 1950s. Because fertility rates had been low two decades before, there were few young adults entering the labour force in the 1950s. Easterlin argues that this limited supply of labour relative to demand pushed up the wages of young adults. Beginning in the 1960s, however, the situation changed. Higher fertility levels in the 1940s and 1950s meant more young adults. Hence the competition for available jobs increased, wages for young adults did not keep pace, and there was substantial youth unemployment. The data in Table 75 indicate that young adults did not share in the prosperity of the 1960s and 1970s to the same extent as their elders. Therefore can the rising prosperity argument still be used in the case of young adults?

I think so. In examining tenure among young adults, it is important to look at more than just their own incomes. One must also consider the financial contributions of parents. These contributions included outright donations of money used as part of the downpayment on a house purchase and loans in lieu of, or to complement, conventional mortgages. In other words, some parents may have used a substantial part of their own wealth to help their grown-up children acquire a first home.

Little information is available on the extent of parental financial support. However, Table 81 provides indirect evidence. In the first row in panel (a), it is estimated that, at an income of $15,000, 16 per cent of young husband-wife couples living alone were owners. The corresponding figure in 1976 was 20 per cent. However, this is for the same nominal income. A $15,000 income in 1971 is roughly equivalent to $22,500 in 1976, given the change in the CPI. In other words, the probability of ownership rose substantially between 1971 and 1976 holding real income constant. How could a larger proportion of families afford to own in 1976? One possible answer is an increased level of parental support in the purchase of a home; at the same level of real income, ownership is more affordable the higher the downpayment.

Table 82 provides additional support for this argument. The next to right column shows the percentage of husband-wife families (with children) who were owners in 1971. The right-hand figure is the percentage predicted using the average income for that cohort in 1971 (inflated to 1980 dollars) and the tenure logit model estimated for 1980. In other words, the right-hand column gives the percentage owning in 1971, assuming that the propensity to own at a given real income had been the same in 1971 as in 1980. The difference between the last two columns indicates the extent of change over the 1970s. Consider the first row in panel (a). Among young three-person units in large urban areas, 28 per cent were owners in 1971 and 56 per cent in 1980. This dramatic increase is only partly attributable to the growth of income. The right-hand column suggests that the ownership propensity would have been 46 per cent in 1971 if the tenure logit model had not changed. Thus sliding along an income-tenure logit curve with the growth in real income accounts for only about two-thirds (or [46-28]/[56-28]) of the increasing propensity of ownership. The remainder is attributable to a shift in the position of the curve itself.

What caused this shift? It might reflect an increased incidence of intergenerational wealth transfers, from parents to children. Also, there may have been a change in the mix of households within this age cohort. The ascendancy of the baby boom generation had, by 1980, pushed up the average age of household heads under 35. Older heads within this age group are more likely to own. As well, house price inflation in the mid-1970s reduced the real cost of ownership relative to renting, hence increasing the attractiveness of owning. Also, the introduction of capital gains taxation in 1972 made the tax-free gains of a principal residence relatively attractive. Finally, the condominium boom of the 1970s and the establishment of several subsidies for low-cost owner-occupied housing opened up the possibility of ownership at incomes that might otherwise have been simply too low.

On net, rising incomes alone did not produce a substantial change in ownership. However, this is not to say that prosperity was without effect. It undoubtedly advanced the date at which a young household could first afford ownership. It also made ownership more affordable to larger families. However, ownership was already high among most husband-wife households and therefore could not increase much. Further, much of the effect of rising prosperity was offset: first by the rise in family size in the 1950s, later by the rise of the nonfamily household and the lone parent. Even in the 1970s, the rapid rise in ownership may be as much attributable to other factors (e.g. the decline in family size) as to rising income.

Expenditure

Next, let us look at income elasticity estimates where housing consumption is measured in terms of housing expenditure. One of the first major cross-sectional studies of this kind is Lee (1968). Using a sample of U.S. households in the early 1960s, Lee regressed rent expenditure, or expected selling price in the case of owners, against the income and demographic characteristics of households.

$$pq = b_0 + b_1 y + \sum_{y=1}^{J} c_j z_j, \tag{33}$$

where the b and c values are estimated coefficients, and z_j is the jth demographic characteristic of the household. Depending on the estimation method employed and the demographic characteristics, Lee finds an income elasticity of between 0.3 and 0.7.

These elasticity estimates are smaller than Muth's. In a sense, this is not surprising. Lee considers only expenditures on the principal residence. Muth's measure of housing consumption, in contrast, includes investment in second homes. Muth's income elasticity of housing demand should therefore be higher, given an income elastic demand for second homes. A second reason for Lee's low income elasticities has to do with tenure. We have already seen that tenure is responsive to income. As incomes go up, some renters switch to ownership. This is embedded in Muth's income elasticity estimates. Lee, however, calculates income elasticities for a given tenure. He shows how sensitive rental expenditure is to income, assuming that the household remains a renter. He does not consider how housing expenditure shifts with a rise in income if there is a concurrent switch in tenure.

One implication of Lee's low income elasticities is that, because the dependent variable is housing expenditure, the share of a consumer's budget spent on housing declines as income rises. On the one hand, this result is commonplace. Most cross-sectional studies of housing consumption find that the expenditures increase less than proportionately with income. However, as noted earlier, the typical budget share allocated to housing did not fall during the postwar period, in spite of rising prosperity. This temporal stability indicates an income elasticity of housing expenditure close to 1.0.

How can Lee's cross-sectional estimates of 0.3 to 0.7 be reconciled with this? We have already suggested two responses: that Lee's estimates use a narrow definition of housing expenditure and that he ignores the effect of income change on tenure. Another possibility is that Lee omits a price variable on the right-hand side of the demand equation. His income elasticity therefore estimates the net effect of income and price variation. Lee's sample includes nonfarm spending units throughout the United States. Since units in large cities tend to have higher incomes and face higher housing prices than do other areas, the true income elasticity is underestimated. In other words, the estimated income elasticity of housing expenditure might well have been larger had Lee taken into account price variation.

Steele (1979) describes a similar analysis of Canadian housing demand based on the Toronto and Montreal subsamples of the 1971 Census public use sample. Her dependent variables are the same as Lee's: rent expenditure or house selling price. Again, this is restricted to the principal residence. The independent variables include several measures of income; age, sex, marital status, labour force status, and education of head; and some indicators of household size. The model fitted is similar to (33).

Steele also finds (186–7) low income elasticities. Her estimates range from 0.2 to 0.5, not unlike those of Lee. This is in spite of the fact that her analysis is restricted to just two cities. The price of a standard quality unit of housing can vary within a metropolitan housing market; however, with a competitive construction sector, such variations should be small. Variations in price are likely more substantial from one city to the next, reflecting such local influences as zoning bylaws, building codes, and climate. It is suggested above that Lee's low estimates might be attributable to a covariation between price and income, given the range of geographic locales considered. Steele's findings make that argument less tenable.

Steele attributes the low elasticities to something else. She argues that income is correlated with other independent variables in the model, e.g. level of education. As a result, part of the effect of income shows up in the coefficients of these other variables. To show the full effect of postwar prosperity, one needs to look at concurrent changes in levels of education, income, and other indicators of prosperity. To look just at income alone is to underestimate the effect of prosperity on housing expenditure.

Why else might the budget share have remained constant over time if the income elasticity was so low? Suppose the demand for housing is given by:

$$h = ay^{b}p^{c}. \tag{34}$$

Note that b and c are the income and price elasticities of demand respectively. This model can be rewritten as follows:

$$ph/y = ay^{b-1}p^{1-c}. \tag{35}$$

The left-hand side is the budget share allocated to housing. Following Lee and Steele, suppose that housing demand is inelastic, i.e. $(b-1) < 0$. As income rises, the budget share would be expected to decline. However, this can be offset by a rise in price if the price elasticity is sufficiently small. In other words, if c is small enough, an increasing price may have kept the budget share of housing from falling even if the income elasticity was low.

In Miron (1984), I use a housing demand model that is based on a Stone-Geary utility function:

$$U = (h-s)^{B}(x-t)^{1-B}: \quad 0 < B < 1,\ h \geq s,\ x \geq t, \tag{36}$$

$$s = a_h + b_h N, \tag{37}$$

$$t = a_x + b_x N, \tag{38}$$

where h is the quantity of housing consumed,
x is the amount of other goods consumed,
B is the relative preference for housing,

s is the "minimum" consumption of housing,
t is the "minimum" consumption of other goods, and
N is household size.

Here, s and t are the consumptions of housing and other goods at which the consumer's utility becomes zero. In other words, it is assumed that the consumer will always purchase at least these minimum amounts (implying further a sufficiently large income). The minimum amount of a good is hypothesized to vary positively with household size.

A demand model is formed by maximizing (36) subject to (37), (38), and a budget constraint. Assuming that other goods are the numeraire, the demand for housing is given by:

$$h = a_h(1-B) + (1-B)b_hN + B(y/p) - Ba_x(1/p) - Bb_x(N/p). \quad (39)$$

Here, p is the price of housing, and y is household income, both taken relative to the price of other goods. This is a linear demand model with four independent variables: N, y/p, $1/p$, and N/p. The parameters of this demand model can be uniquely identified, given estimates of the intercept and slope coefficients of (39).

This demand model was estimated using a sample of owner and renter spending units from across Canada in 1978. Housing consumption, h, was calculated as housing expenditure discounted by price. As with Lee and Steele, housing expenditure was defined to be the annual rent paid by tenants, or the dwelling market value for an owner-occupier. To calculate the price discount, Canada was divided into 20 areas by geographic locale and city size. A weighted average price was calculated for a standardized mix of housing units by quality level, and this was divided into expenditure to give housing consumption. The price of housing, p, on the right-hand side of (39) was estimated using the price discounts.[23] Income is defined to be the current total income of the household. The model thus differs from Lee and Steele in that the dependent variable, expenditure, is price-discounted; price is included as an independent variable; and current, rather than permanent, income is used.

Price and income elasticities were estimated for both renters and owners. In addition, the sample was disaggregated by six types of households, and separate price and income elasticities were estimated for each. The elasticity estimates are presented in Table 83. Consider first the income elasticities. These are close to 0.3 in every case. The differences between owners and renters are small. The variations are only slightly larger among demographic groups. They are consistent with the argument that demographic changes did not affect the response of aggregate housing demand to rising income. For example, the rising incidence of persons

TABLE 83
Income and price elasticity of demand by tenure and type of spending unit, Canada, 1978

	*Elasticity of demand**			
	Income†		*Price*‡	
Type of spending unit	*Owners*	*Renters*	*Owners*	*Renters*
All spending units	0.33	0.38	−0.27	−0.14
Units with children under 18				
Husband-wife units				
Youngest child under 5	0.33	0.37	−0.28	−0.27
Youngest child over 5	0.34	0.44	−0.31	−0.10
Lone-parent units	0.16	0.15	−0.36	−0.31
Units without children under 18	0.27	0.36	−0.23	−0.11
Head under 35				
Head 35–54	0.41	0.53	−0.32	0.01
Head 55 or older	0.25	0.26	−0.13	−0.20

SOURCE: Statistics Canada, Survey of Family Expenditures in 1978 micro data tape. Calculations by J.R. Miron

* Elasticity estimates based on Stone-Geary demand model. See text for more details. Model estimates derived using weighted least squares. See Miron (1984) for details of sample and estimation method.

† Current total household income used as income measure.

‡ Measured as described in text

living alone, or the decline of the large family, did not lead to a rise in overall income elasticity, as both groups had an income elasticity near 0.3.

As with Lee and Steele, these income elasticities have to be interpreted carefully. They refer only to a restricted category of housing expenditure and ignore the effect of income on tenure. Perhaps because of these similarities, the elasticity estimates are close to those obtained by Lee and Steele.

In equation (35), the decline in the budget share with rising incomes can be offset by a rising price of housing if the price elasticity of demand is sufficiently low. In Table 83, the estimated price elasticities are indeed small. It is thus possible that a rising price for housing, combined with inelastic demand (i.e. low price elasticity), kept the budget share from declining in a period of growing prosperity. The fly in this ointment is the postwar behaviour of SCHI. As noted in Table 69, this housing price index increased at about the same rate at the overall CPI between 1946 and 1981. In other words, the price of housing just kept pace with the the prices of other goods. At the same time, though, real incomes rose. Rising real income, combined with low income elasticty, should have depressed the budget share of housing. This should not have been counteracted by low price elasticity if housing prices did not increase relative to the prices of other goods. Thus SCHI negates the possibility that the budget share remained constant because of a rising relative price for housing.

So this discussion terminates in a quandary. Cross-sectional evidence suggests that the income and price elasticities of demand for dwelling quality are low. Given the rapid growth of real income, and that housing prices have just kept pace with other consumer prices, the budget share of housing should have declined over the postwar period. However, Table 69 suggests that it has remained constant and may even have increased during the 1970s.

Why? In part, it may be because cross-sectional studies use a restricted definition of housing expenditures and ignore the effect of income on tenure change. Housing expenditures associated with second dwellings – such as cottages or chalets – are ignored. Also ignored are certain costs of household operation such as furniture, appliances, and floor coverings. Such expenditures are included, however, in the time series data in Table 69. Over the postwar period, there was a rapid expansion in the stock of cottages and other kinds of second homes (although measures of this are not readily available). I think that there was also a sharp escalation in the amounts spent on household furnishings and operation. In other words, consumers used their postwar prosperity partly to improve their principal residence but also to outfit extensively their homes and to acquire second dwellings.[24]

CONCLUSIONS

To what extent did increasing prosperity help shape postwar Canadian housing demand? Chapter six considers the role of rising prosperity on living arrangements. There, evidence is presented that the effect of prosperity was mainly on the formation of nonfamily households. Families tended to maintain a dwelling regardless of income, and the taking in of lodgers had little to do with affluence. Given that, for every new household on net, there must be an additional occupied dwelling, this effect of prosperity on housing demand was substantial. Hickman's U.S. data suggest that the income elasticity of demand for new dwellings – remember that some of the demand can also be accommodated by conversions of the existing housing stock – is about 0.2.

This chapter considers how prosperity shaped the kinds of dwelling units occupied. It focuses on the effect of income on tenure choice and dwelling quality. The estimation of a housing demand model is fraught with methodological and conceptual problems. The important points to note are that we generally observe only housing expenditure, not "price";

that there are several ways of decomposing expenditure into price and quality components; that there are several ways of measuring consumer income; and that there are a number of different approaches to the modelling of demand. This makes it difficult to draw generalizations from studies that use different methods and concepts.

With this caveat in mind, let us consider some measures of the income and price elasticity of demand for dwelling quality. Most studies define the quality of a dwelling as the selling price, or monthly expenditure, discounted by the price of a standard-quality unit. Using time series data, Muth found an income elasticity of new dwelling quality of 0.9. Hickman obtained a value of 0.6, but this may be an underestimate because of multicollinearity. In cross-sectional studies in both the United States and Canada, values of 0.3 to 0.7 are frequently reported. However, these are based on a narrowly defined category of housing expenditures for principal residences only and ignore tenure change. In other words, none of these results necessarily contradicts the unit elasticity implicit in the constant budget share allocated to a broadly defined category of housing expenditure as evidenced over the postwar period.

Estimates of price elasticities vary more widely. Muth finds an elasticity of −0.9 using time series data. My own cross-sectional estimates are much lower, nothing much larger than −0.3. However, this is not surprising. Price elasticities are sensitive to the way in which housing expenditure is decomposed into price and quantity components.

In any event, the exact magnitude of the price elasticity may not be important. If SCHI correctly reflects housing prices, it suggests that these have just kept pace with other prices over the postwar period. Thus the effect of house price change relative to other price changes may well have been minimal. The dominant economic change was the rise in real income, rather than a shift in the relative price of housing.

How important was the rise in real income in shaping postwar housing demand? One way to answer this is to contrast its effect with that of demographic change. We have already extensively noted the rise of the one-person household, itself partly a response to rising real incomes. This decline in average household size itself led to the construction of smaller, less expensive dwellings. Hickman's analysis suggests that the decline in average household size may have largely offset the effect of rising income on typical dwelling quality. Put another way, the effect of prosperity on the housing stock would have been more evident had there not been an attendant decline in household size.

CHAPTER NINE

Public Policy and Housing Supply

During the postwar period, as in earlier times, Canada's housing stock was predominantly privately owned. Over one-half of all Canadian households owned their own dwellings. In addition, much of the rented housing stock was owned by individuals or corporations (nonprofit and profit) in the private sector. While there was a substantial increase in government-owned/operated housing, it formed only a small part of the total housing stock. As of 1975, for example, there were only about 140,000 rental public housing units, roughly 2 per cent of the total housing stock.

This does not mean that governments played a limited role in development of the housing stock. For a commodity that was predominantly privately traded, the housing stock was significantly affected by government activity: in both quantity and quality. To see this, one needs to look at the effects of governments more broadly than simply in terms of public housing units. Through regulation and subsidy, governments have substantially affected the provision of private-sector housing.

This chapter explores the various effects that governments in Canada had on development of the postwar housing stock. Such a broad topic cannot be adequately covered in this short space. One should also refer to Central Mortgage and Housing Corporation (1971), L.B. Smith (1971, 67-88), Dennis and Fish (1972), Saywell (1975, 189-216), Canadian Council on Social Development (1977), Task Force on CMHC (1979), Fallis (1980, 9-37, 143-60), Rose (1980), L.B. Smith (1981), and Dowler (1983). These reviews detail the development of postwar housing policy in Canada. Complementary analyses of prewar housing policies are found in Firestone (1951, 107-31) and Jones (1978). The intent here is not to provide a thorough review of Canadian housing policies, so much as an impression of the extent of public-sector effects.

Constitutionally, Canada has two levels of government, federal and provincial, each with its own areas of jurisdiction. A third level of government, municipal or local, is a creation of the provinces; it is not given explicit powers in the constitution. Although housing is not specifically mentioned in the constitution, provinces are given control of management and sale of public lands, property and civil rights, local works and undertakings, and matters of a local or private nature. At the same time, residual powers, i.e. those not specifically allocated to either level, are assigned to the federal government under the peace, order, and good government clause. In practice, both levels actively developed legislation and programs that affected the development of the housing stock. This complicates discussion of the effects of governments; a complete discussion necessarily must consider each of the ten provinces. Further, each province passed enabling legislation that allowed municipalities to undertake some planning and regulatory functions related to housing. The exact mix of legislation, regulation, and policies under which the housing stock developed thus varied from one area to the next. A thorough review of these factors in combination is beyond the scope of this chapter. Instead, the chapter merely sketches some important apects, focusing mainly on the federal government and its role.

The remainder of the chapter is divided into two parts. The first reviews how governments in Canada became involved in housing. Over the years, governments have faced different kinds of problems, some directly concerned with housing, others only indirectly. Programs and policies were developed that affected the housing stock. As the problems faced by governments changed, so did the policies and programs. New policies emerged, and old ones disappeared. This discussion helps us to understand the great variety of housing programs and policies introduced in the postwar period. The second part of the chapter describes some of these programs and policies and assesses their effects on the housing stock.

RATIONALES FOR PUBLIC-SECTOR INVOLVEMENT

The first modern instance of direct federal involvement in Canadian housing was a $25-million loan to the provinces initiated in December 1918. The loans were used for the construction of new dwellings. Authority for the loans was taken from the War Measures Act, which gave the federal government broad emergency powers.[1] Jones (1978, 13) indicates that the federal government justified these loans by arguing the national importance of housing in its relationship to the health and well-being of all Canadians, the welfare of returned soldiers and their families, and the employment to be generated during postwar reconstruction. Constitutionally, these could be perceived as federal matters. Jones also suggests (p. 13) that government concerns about labour unrest, agricultural radicalism, and the Bolshevik revolution also increased its receptiveness to programs, such as housing construction, that would "stabilize the social situation."

Over the years, the federal government's rationales for its housing programs changed. After the Second World War, the federal government again initiated a program to help house returning veterans and to ease structural unemployment as the economy shifted from a wartime footing. However, by this time, federal activity in housing was much more widespread. During the Depression, the government adopted some of the New Deal legislation enacted in the United States. In particular, the Dominion Housing Act of 1935 and the Home Improvements Loan Guarantee Act of 1937 were intended to bolster renovation and new housing construction. These acts were seen primarily as ways of increasing employment in the construction sector. Unlike the immediate postwar programs, however, they were not seen to be simply short-term relief programs. The National Housing Act (NHA) of 1938 appeared as the national economy had begun to recover. It manifested an unprecedented broadening of federal interests. It continued the Dominion Housing Act's emphasis on job creation. However, in addition, it created for the first time a direct federal role in the production of housing geared to low-income households. The 1944 amendments to the NHA involved the federal government in providing aid for slum clearance and urban renewal. Thus, by 1946, the federal role in housing had grown considerably. Its set of concerns had expanded from short-term unemployment relief to include social housing issues. In particular, it was increasingly based on the view that housing for the poor was inadequate and that government action was necessary to redress the situation.

There was an increased emphasis over time on this use of housing as a tool of social policy. The 1964 amendments to the NHA emphasized explicit programs for urban renewal and low-income housing. In 1973, further amendments to the NHA were introduced by the minister responsible who described "good housing at a reasonable cost" as a "social right."[2] Increasingly, the distinctions between federal and provincial objectives in housing policy became blurred.

Additional federal motives became apparent. Part of the attraction of NHA mortgages was that they were linked to small downpayments. In 1949, downpayments were reduced even further in an effort to bolster employment in the construction sector. However, at the outset of the Korean War, the government faced another problem, inflation. As part of an effort to curb private-sector demands everywhere, the minimum downpayment on an NHA mortgage was raised in 1951.

Finally, there was a considerable broadening of the federal interest in urban development generally, beginning in the late 1960s. In 1968, the government established a Task Force on Housing and Urban Development. In its report, the task force urged the federal government to become much more involved in shaping urban development.

In fact, the federal government did not immediately implement many of the recommendations of the task force. It did create, however, in 1972 a Ministry of State for Urban Affairs (disbanded later in the decade). Broadly, the purposes of MSUA were to advise the government of the consequences of its various policies and programs for urban development in Canada and to co-ordinate the actions of various federal departments so that certain desirable urban policy objectives could be achieved. Although still largely independent, CMHC reported to the minister of state for urban affairs, and the corporation was seen by MSUA as a principal federal actor in shaping urban development. MSUA had a relatively short life. While it existed, it symbolized federal interest in a broader role in urban planning and development. In this sense, housing was seen as just one aspect of a broad set of social concerns associated with urban growth.

Let us now contrast these federal rationales with those of the provinces and local governments. Under Canada's constitution, local governments are creations of the provinces. In addition, the provinces have jurisdiction over land, property, and matters of a local or private nature. These powers were used to shape the growth of the housing stock. By 1981, most or all of the provinces had legislation – in areas such as public housing, local government, urban and regional planning, land subdivision, and environmental quality – that helped shape the housing stock. Some aspects of this legislation are discussed later in this chapter. For the moment, let us consider simply the broad rationales in establishing these acts.

We have already considered one. With increasing urbanization, it became clear to most provinces, in some cases with the prodding of local governments, that some important social and economic problems resulted from inadequacies in the housing stock and the way that urban areas were developing. Among these were the health problems associated with primitive sewage and garbage disposal systems; untreated drinking water; slum housing and overcrowding; inferior housing design and construction, including poor ventilation and inadequate access to sunlight; air, water, and noise pollution; and traffic congestion. Other concerns were for economic, social, and crime problems arising from the divergence between "haves" and "have-nots."

The remainder of this chapter examines some of the public programs and policies that shaped the postwar housing stock. For discussion purposes, eight major categories of public involvement are identified: in mortgage financing, in other subsidies to owner-occupiers, in public housing construction, in other rental subsidy programs, in tax subsidies to landlords, in the handling of owner equity under the Income Tax Act, in land subdivision controls, and in zoning. These categories are not exhaustive but do illustrate the important and varied effects of the public sector.

MORTGAGE FINANCING FOR OWNER-OCCUPIER HOUSEHOLDS

Beginning with the loan plan of 1918, the federal government acted to alter the typical terms of mortgage financing in Canada. Gradually, this evolved into a deliberate policy of encouraging "easy money" for housing, which effectively resulted in more, inexpensive housing. To a large extent, it may also have encouraged owner occupancy.

The 1918 loan scheme antedated the "easy money" policy. However, it was a prototype for future legislation. In it, the federal government made available $25 million to the provinces on a per capita basis.[3] The loans were made at an annual interest rate of 5 per cent with a 20- to 35-year term. At the time, these were low interest rates and long terms compared with mortgages available from private sources. Conditions were imposed on the provinces, including a maximum selling price of dwellings financed by these loans. In addition, the provinces had to contribute $1 for every $3 of federal monies in making housing loans. Further, the loans made by the provinces had to be used exclusively for construction of new owner-occupied dwellings. In effect, the program made new, moderately priced, owner-occupier housing available with below-market interest rates and a generous repayment period. About 6,000 dwellings were constructed under this scheme.

The Dominion Housing Act of 1935 was the first major, long-term move by the federal government into the mortgage market. This act authorized the issuing of joint 80 per cent first mortgages to new home buyers. The mortgage was issued by a lending institution but held jointly by the lender and the government. The federal government advanced 20 per cent of the cost of construction (or appraised value) to the lender at 3 per cent interest. The first mortgage itself was issued by the lender at 5 per cent with a 20-year amortization. The low downpayment, low interest rate, and long term were unmatched in the private mortgage market. Firestone (1951, 481) reports, for example, that most private mortgages at that time were for no more than 60 per cent of dwelling value and had terms of only 5 to 10 years. The act was soon replaced by the National Housing Act of 1938; only about 4,900 dwellings were completed under it.

The Veterans' Land Act of 1942 also included provisions related to mortgage financing. Under it, veterans purchasing housing constructed by the Department of Veterans Affairs could obtain mortgages at 3.5 per cent interest amortized over 25 years with a downpayment of only 10 per cent. The act also provided for cash grants. Firestone (1951, 489) reports that over 10,000 dwellings were started under this program between 1946 and 1949. The program was terminated in 1975, with about 40,000 dwelling starts being assisted in total.

In the National Housing Act of 1938, the above-described provisions of the Dominion Housing Act were continued. In addition, for low-cost homes, the first-mortgage limit was raised to 90 per cent, further lowering the required downpayment. The 1944 rewriting of the NHA reduced the interest rate to 4.5 per cent and increased the maximum amortization period to 30 years. Saywell (1975, 203) reports that downpayment requirements of joint loans were further reduced in 1949 to stimulate housing construction but raised again in 1951 as an anti-inflationary measure.

The 1944 act also introduced direct lending; the newly created Central Mortgage and Housing Corporation (CMHC) was authorized to make mortgage loans available (on the same beneficial terms as joint loans) in geographic areas not well served by conventional lenders. This "residual lender" role was not substantial, however; Firestone (1951, 496–7) reports that only 781 dwellings were so financed between 1947 and 1949.

In 1954, the National Housing Act was again rewritten. Joint loans were replaced by insured loans. The approach was different, but the results were similar. The insurance program placed a premium on an approved mortgage. CMHC collected the premiums and used this pool of funds to pay off default losses. The scheme was intended to be actuarily sound, the net cost of the insurance program to the government being negligible. In effect, the program reduced the risks of mortgage default, thereby encouraging more lenders to participate in the market and at lower interest rates than otherwise possible. To the federal government, the insurance program was attractive; it offered assistance to home buyers at a minimal expense to itself. At the same time, eligibility requirements limited the overall cost of the program. The government set maximums for the loan-to-value amount, permissable loan, amortization period, interest rate, and ratio of gross debt service to income. There were also regulations governing the kinds of dwellings eligible for an insured mortgage; over the years, they gradually came to require that the dwelling meet the requirements of the National Building Code (a model code developed by the National Research Council) and other design criteria.

The principal problem in relying on an insurance scheme to promote housing is that it does not guarantee a supply of mortgage funds. Insurance is an inducement to make mortgage money available; it does not compel the capital market to produce funds. The 1954 NHA therefore also included provisions to increase the supply of mortgage money. Previously, chartered banks and caisses populaires had been excluded from NHA lending. The new act opened the field to them. In addition, CMHC was given the authority to buy and sell insured mortgages, which helped to increase liquidity in the market.

The mortgage insurance program worked well for a few years. Beginning in 1956, however, the volume of insured mortgages dropped sharply when the interest rate ceiling on insured mortgages became unattractive. Although the maximum NHA interest rate was raised to 6 per cent in 1957, this did not substantially increase lending.[4] In midyear, the government widened its residual lending to cover all Canada. Residual lending accounted for almost one-third of all residential mortgages in 1957.[5] In 1957, 17,000 dwellings were so financed, rising to about 27,000 units in 1958 and 1959.[6] Needless to say, the direct lending program put a strain on government finances. It was subsequently replaced by a Small Homes Loans Programme, which restricted direct lending to houses of under 98.5m^2 (or 1,060 ft^2). Fallis (1980, 15) estimates that between 1957 and 1968, 16 per cent of all new dwelling units in Canada were financed under the residual lending program.

The 1960s saw another significant shift in government policy regarding mortgages. The government reduced the minimum term on insured mortgages to five years. In effect, this made possible a difference between the "term" of a mortgage (the length of time after which a mortgage becomes due) and the "amortization period" (the date at which the remaining balance of the mortgage has dropped to zero). In other words, a borrower could obtain a mortgage amortized over 25 years but due in 5 years; at the end of the 5 years, the borrower either repaid the remaining principal outstanding or obtained a rollover mortgage to continue financing. This scheme made available money for mortgages that a lender might be unwilling to put into a longer, fixed-term loan.[7]

In 1973, the federal government introduced the Assisted Home Ownership Plan (AHOP), designed to assist home owners by reducing the cost of mortgage financing. In its original form, it provided 95 per cent first mortgages, amortized over 35 years at 10 per cent interest, to families (with dependent children) of modest income and provided subsidies to reduce mortgage payments to a minimum interest rate of 8 per cent or 25 per cent of family income. A supplementary grant (at first $300, later $600) was available if, at 8 per cent interest, the mortgage payments still exceeded 25 per cent of income. CMHC specified maximum qualifying family incomes and house prices, and the program was restricted to first-time home buyers. In 1974, AHOP was further restricted to newly constructed dwellings. AHOP existed in this form from 1973 through 1975. Approximately 40,000 households were assisted. The cash flow associated with direct loans made by CMHC under this program was in excess of $1 billion.

In late 1975, AHOP was substantially modified. The interest reduction grants were replaced with loans in the form of second mortgages. The second mortgage was provided interest-free for five years. Advances under the second mortgage followed a sliding scale, the effect of which was to increase annually the monthly payment required over the first five years. In effect, this resulted in a graduated-payment mortgage. For households expecting substantial income gains in the near future (e.g. young families), this scheme made ownership more affordable. This second version of AHOP was terminated in mid-1978. About 93,000 households secured loans under this version of AHOP.

In general, the federal government acted over the postwar period to make "easy money" available to home (especially new home) purchasers. Through direct lending, mortgage insurance, and various interest reduction schemes, it attempted to provide longer amortization periods, lower interest rates, and lower downpayments than the private sector had been willing to provide. The purpose was to make owner occupancy more affordable to households of modest income or savings. Certainly, it is evident that the federal government, acting principally through CMHC, was substantially involved. In a review of its activities between 1946 and 1970, Central Mortgage and Housing Corporation (1971, 6) concludes: "Since 1946, year after year, between one-half and one-third of all housing units built in Canada has [sic] been financed through NHA arrangements . . . Of the three million units built in this 25-year period, more than 1,000,000 have come through NHA channels; and of this one million about half has been financed entirely with government funds." This estimate covers the period up to 1970. L.B. Smith (1981, 342) estimates that 45 per cent of total housing starts between 1970 and 1978 were made under NHA provisions. Although these estimates include all forms of NHA activity, well over one-half of NHA-related starts were for owner-occupier units.

However, this does not necessarily mean that NHA had a large net effect. NHA did lower mortgage costs, or make mortgage money more available, for many households. However, some of them would have become owners even in the absence of NHA. Others might have also become owners anyway, but later in their life cycle. By reducing downpayments and stretching out interest payments, NHA allowed them to become owners at a younger age. The net effect of NHA can thus be measured in two ways: as the number of households that became owners that might not have otherwise and/or as the number of household-years by which ownership increased because of the presence of NHA. Further, government borrowing in the capital market to finance various mortgage programs cut into the capital available (and/or an increase in the cost of such capital) for other purposes, including private mortgages. In other words, part of NHA activity was "purchased" at the expense of private mortgage lending.

I am not aware of an overall estimate of the net effect of NHA activity on the postwar housing stock that takes these considerations into account. There are, however, strands of evidence taken from different sources that appear to tell a consistent story. In a U.S. study, Murray (1983) looked at the effects of subsidized owner-occupied housing starts on the volume of unsubsidized starts in the 1960s and 1970s. In his model, the demand for unsubsidized starts is positively related to real income and negatively related to the costs of construction and financing and the volume of unsubsidized starts (for each of low-income and moderate-income housing units). The supply of unsubsidized starts is positively related to the flow of mortgage-eligible funds and the mortgage-prime interest rate spread and negatively related to the number of subsidized starts (for each of conventionally financed and government-financed subsidized units).[8] Numerical simulations with this model indicate that a subsidized housing start that was conventionally financed (the typical mode in moderate-income programs) largely eliminated an unsubsidized start. A subsidized start that was government financed (the typical mode for low-income programs) did, however, have a positive aggregate effect on housing construction. Murray (1983, 590) concludes: "empirical evidence from the period 1961-1977 indicates that most of the effect of subsidized starts on the housing stock was offset by the displacement of unsubsidized starts. I estimate that one group of programs [conventionally financed] had no measurable effect on the stock of housing, and that more than half the effect of [government financed] starts during the period was lost to displacement."

Lithwick (1978a) attempted to measure the net effect of Canada's Assisted Home Ownership Plan (AHOP). He notes (p. 136) that AHOP "was directly responsible for approximately 22,000 new housing starts in 1976." Lithwick used the Bank of Canada's econometric model to predict the level of investment in new housing that might have occurred in the absence of AHOP in the same year. He finds (p. 134) that the net contribution of AHOP to the total housing stock was only about 10,000 units. In other words, about one-half of the AHOP units started would have been built even without the program. This finding is similar to Murray's. AHOP was aimed at low-income households; it was an attempt to attract households that otherwise might not have been able to afford ownership. In its effect, it was similar to that of the low-income subsidies reported by Murray. L.B. Smith (1981, 345) suggests that the net effect of AHOP may be even smaller still, perhaps as little as 13 per cent to 27 per cent of the gross number of dwellings completed under the program.

Finally, L.B. Smith (1974, 140-51) has examined the short-run effect of CMHC's residual lending program on the number of conventional and NHA housing starts. This program includes lending for both owner-occupied and rental housing. For a period covering the 1950s and 1960s, he found (p. 150) that, in any two-year period, "[a] million dollars in constant 1957 dollars of CMHC lending . . . generate[d] an additional 79.9 NHA financed housing starts . . . but . . . cause[d] conventionally financed housing starts to be reduced by 18.2 units as demand that would otherwise be satisfied in the conventional market is satisfied by CMHC direct loans." The number of dwellings contructed overstates the net effect of NHA residual lending by about 30 per cent. This is a larger net impact than is reported by either Lithwick or Murray. It is, however, consistent with the notion of CMHC as lender of last resort. Because CMHC tended to make direct mortgage loans only where there was no other apparent source of funds, it should not be surprising that few of these dwellings would have been constructed in the absence of residual lending programs.

In conclusion, the net effect of these mortgage financing schemes was somewhat smaller than the number of dwellings constructed under their auspices. In general, the net effect was likely very small in the case of programs oriented primarily toward moderate-income households and somewhat larger where low-income groups were the target or where alternative financing was unavailable.

OTHER SUBSIDIES TO HOME OWNERS

At various times, the Canadian federal government also undertook other programs to assist home owners. One of the earlierst was the Home Improvements Loan Guarantee Act of 1937. This was a loan insurance plan for improvements or extensions to residential property.[9] Approved lenders could make loans of up to $2,000 for a single house (more on multi-unit structures) but were limited to 3.25 per cent interest on a one-year term (there were higher rates for longer terms). The program was terminated in 1940. About 126,000 loans were handled under the act, roughly one loan for every 20 dwellings in Canada. Just under 4,000 of these loans were used for conversions, thus adding to the stock of rental dwellings.

During the Second World War, two new renovation programs were introduced. One was the Home Extension Program of 1942. Providing loan guarantees for home extension and conversion, it lasted only until 1944 and was not widely used. The other was the Home Conversion Plan of 1943. Under it, the government leased large dwellings in major urban centres for conversion into multiple apartments. Firestone (1951, 491) reports that about 2,100 conversions were carried out under this program. In this sense, the program contributed to the total stock of dwellings and resulted in relatively more rental units. There is little information on postwar deconversion; presumably many of these conversions remained in the rental stock after the war.

Late in the war, the federal government introduced programs to assist in postwar demobilization. One such program was the Integrated Housing Plan.[10] In it, the government entered into agreements under which a builder would construct specified dwellings. The government agreed to purchase the dwelling at a predetermined price should the builder not be able to sell it within one year. The government then sold such acquired units to returning veterans. In effect, the plan guaranteed a minimum selling price to the builder, who then was better able to secure interim financing. In this respect, the plan encouraged new dwelling construction, especially of owner-occupier units.

Since 1955, the NHA has also provided for home improvement loans. The federal subsidy took the form of a guarantee on such loans. Between 1955 and 1981, over 450,000 dwellings received such loans. With the introduction of RRAP (discussed below), use of the home improvement loan program declined.

Another federal subsidy program was the Winter Works program of the early 1960s. This program provided cash grants for dwellings constructed during the winter months. It was conceived as a program to reduce seasonal unemployment in the construction industry but also provided an incentive to home ownership.

A 1973 amendment to NHA provided for the Residential Rehabilitation Assistance Program (RRAP). RRAP's purpose was to improve the existing low-income housing stock. Under RRAP, loans up to $5,000 were made to home owners of modest income in selected neighbourhoods.[11] The owner had to agree to remain in the dwelling for the term of the loan. Up to one-half of the loan amount was forgiven, depending on the household's income. Between 1974 and 1978, RRAP loans were made to 55,634 dwellings.[12]

Finally, there were programs to upgrade specific aspects of housing, notably those related to energy use. Two examples were the home insulation programs (known as CHIP and HIP) and the Canada Oil Substitution Program.

These programs provided subsidies to home owners. Their effects on Canada's postwar housing stock may have taken several different forms.

Some of them, such as the Home Conversion Plan and the Winter Works scheme, may have increased the aggregate housing stock. Others may have induced households to stay on in a dwelling for which repairs would otherwise be uneconomic, thereby prolonging its life. For still others, the primary effect may have been to assist households to remain owners.

Although these programs were widely used, the net effects on the housing stock are not clear. Presumably, some amount of maintenance, repair, and other such work would have been undertaken by the beneficiaries in any event. In other cases, the home owner might have sold the dwelling to someone who could have afforded to do the necessary work.

INCOME TAX AND HOME OWNERSHIP

Canada's personal income tax system contains several subsidies for home owners. The federal legislation that covered most of the postwar period was replaced in 1971. For ease of reference, let us call the 1971 legislation the "new" act, and its predecessor the "old" act.

The old and new acts differed in their treatment of a capital gain. Both acts distinguished between regular income and capital gain income. Crudely put, a capital gain is an amount arising from a transaction that was not part of one's normal livelihood. The net gain on the sale of a dwelling was treated as a capital gain, for example, provided that the vendor was not in the business of buying and selling dwellings. Under the old act, a capital gain was not taxable. Under the new act, one-half of the capital gain was taxable, an important exeption being the taxpayer's principal residence. Thus, under both acts, the household was exempt from tax on the disposition of its principal residence. In itself, this made home ownership more attractive to those seeking reduced tax exposure. The new act further emphasized this by taxing other capital gains, such as those on the sale of corporate shares. In effect, the new act made home ownership a more attractive investment for capital gain.

That both acts exempted the imputed return on owner equity created another incentive (or subsidy) for home ownership. Consider, for example two taxpayers, A and B, each with $100,000 in accumulated savings and an annual salary income of $30,000. Suppose A invests the $100,000 at 10 per cent. On the $10,000 investment income produced, and $30,000 in regular income, suppose A pays (at a 30 per cent average tax rate) $12,000 in tax. Further, suppose A rents an apartment for $10,000 per year (including fuel, utilities, and other charges), giving an income, net of taxes and shelter, of $18,000. Suppose B uses the $100,000 to buy, at the beginning of the year, a dwelling similar to A's apartment. The dwelling is sold at the end of the year for a net gain (after property taxes, fuel, utilities, and other charges) of $1,000. B's tax (again at 30 per cent average) is $9,000; remember the gain is tax-free. B's net income for the year is thus $22,000 net of tax and shelter cost. In fact, B would have to suffer a net loss on the sale of his property of $3,000 before his situation is equivalent to that of A. It is in this sense that an owner-occupier receives an implicit tax subsidy.

The dollar value of such subsidies is substantial.[13] In 1979, the direct subsidies provided under NHA totalled just under $1 billion. This includes the assisted ownership programs and the implicit subsidy in NHA insurance fees discussed previously. The exclusion of imputed rent on owner equity has been estimated to represent an additional $1.75 billion in lost tax revenue. The lack of capital gains taxation on principal residences is estimated to represent a further tax loss of $3 billion.

The new act also provided for a Registered Home Ownership Savings Plan (RHOSP). Under the plan, lending institutions could offer savings accounts into which taxpayers (who did not already own a dwelling and had not previously had an RHOSP) could put up to $1,000 annually up to a limit of $10,000. The taxpayer could deduct each annual contribution from that year's income for tax purposes. Further, if the RHOSP account were later closed and the money (including any income earned on this money while in the RHOSP) used to purchase an owner-occupied dwelling (including, at certain times, some furnishings), the taxpayer did not have to count the money as part of his or her income in that year for tax purposes. In other words, the taxpayer could get $10,000 plus any accumulated interest as tax-free income to assist in the purchase of a home. In effect, this reduced the "price" to renters of switching to ownership by allowing them to use, toward that purchase, money that would otherwise have been paid in tax. However, RHOSPs were not widely used. It has been estimated that in 1979, for example, the RHOSP program represented a tax loss of only $95 million, small relative to other implicit subsidies for home owners.

I am not aware of any attempt to measure the changing effect of Canada's income tax legislation on home ownership. However, Rosen and Rosen (1980) describe a relevant U.S. analysis. The U.S. experience was somewhat different because, unlike Canada, federal tax law permits the deduction of property taxes and mortgage interest. However, as in Canada, there is no taxation of the imputed return on owner equity. Rosen and Rosen estimate a model of tenure choice in the United States using aggregate time series data from 1949 to 1974. The dependent variable is the proportion of households that were home owners in a given year. The independent variables include income, credit availability, and the ratio of the net price of ownership to renting.[14] The net price of ownership is the aftertax cost of housing. Between 1949 and 1974, the percentage of

U.S. households that were home owners rose from 48 per cent to 64 per cent. From their model, Rosen and Rosen estimate that taxation of the imputed return on owner equity and the nondeductibility of property taxes and mortgage interest would have reduced the incidence of ownership to 60 per cent in 1974. In other words, the elimination of certain tax breaks for owners would reduce the incidence of home ownership only modestly. In Canada, where there are fewer tax breaks, the impact on ownership would presumably be smaller still.

PUBLIC HOUSING CONSTRUCTION

In this chapter, "public housing" refers to housing owned and/or operated by a government agency for the purpose of providing accommodation to individuals and families of low income. Public housing has traditionally been one of the ways that society has attempted to house people who could not otherwise afford decent housing. Other types of public assistance to renters are discussed later in the chapter.

The first explicit recognition of social housing as an ongoing concern of the federal government was in the National Housing Act of 1938. Part II of that act provided for loans at 2 per cent interest to local housing authorities for the construction of low-rent housing. This provision was never implemented, however.[15] The next instance of such legislation was not until 1949, when amendments to NHA allowed the federal government to enter into a 75 per cent–25 per cent sharing of costs with any province, for land assembly and building. Operating deficits were to be shared on the same basis. This again was not a widely used program. By 1955, only 3,000 low-rent dwellings had been constructed; by 1964, the accumulated total was only 12,000 units.[16] The Canadian Council on Social Development (1977, 2–3) comments: "In part, the new provision was necessitated by the federal government's announced intention to vacate its direct lending role with respect to rental housing. The level of participation required of the the provinces was deliberately established at a level designed to keep down the number of public housing projects."

The 1964 amendments to NHA marked a new turn in federal housing policy. In section 35A, the old joint loan program was continued. However, a new section, 35D, allowed the federal government to make 90 per cent mortgage loans, for terms of up to 50 years, for provincially initiated and owned low-rental projects. Rents in the housing projects were geared to income. Operating losses were to be shared equally with the province. In effect, the federal government agreed to assume a larger portion of the cost of public housing under section 35D. Between 1965 and 1981, an additional 25,000 rental units were constructed under the old joint loan program (section 35A).[17] Under the new section 35D, however, over 140,000 new units were created, mostly between the late 1960s the and mid-1970s.[18] By the end of the 1970s, the extensive construction of public housing had come to an end, replaced, in part, by a variety of other forms of assisted rental housing. Nonetheless, public housing (under sections 35A and 35D) formed roughly 5 per cent of all new dwelling completions in Canada between 1965 and 1980.[19]

This is a much smaller figure than that offered above for NHA support of owner-occupied dwellings. However, many NHA-financed, owner-occupied dwellings might have been constructed even in the absence of NHA. The same may not be true for public housing. Before 1964, the provinces had not shown much willingness to bear the substantial costs of constructing public housing. And the private sector did not seem capable of producing adequate new housing at prices affordable to low-income households. In this sense, few if any of the 165,000 dwellings created under sections 35A and 35D might have been built in the absence of NHA. The relative effects of NHA on the owner-occupied and low-rent stocks may, therefore, not be as disparate as might first appear.

At the same time, the assistance given to public housing was not without cost to other kinds of housing. As with other NHA programs, the federal government itself had to borrow in order to finance its public housing program. Some of the financing for public housing thus represents money that might otherwise have gone into private mortgages. In this sense, the large amount of public housing constructed in the late 1960s and the 1970s may well have had a negative effect on the construction of housing in the private sector.

Also, public housing construction may have altered the process of filtering in the housing market. It is commonly argued that low-income households, if left to the private sector, are allocated to older, dilapidated housing. Because of building-code, minimum-size, and other restrictions, new housing simply cannot be provided at prices affordable to the poor. With time, all housing requires maintenance and repair work. If insufficient work is carried out, the quality of the dwelling declines, and its market price slumps. Some owners will, of course, spend enough to maintain, or enhance, their property values. However, given a good supply of new high-quality dwellings, the movement of higher-income consumers from older to newer units will reduce the demand for older units. For some owners, it will be rational to let their properties deteriorate, eventually finding a new niche servicing low-income consumers. By providing an alternative, public housing may have hastened the demolition of older housing. With reduced demand for private-sector low-quality housing, presumably more landlords see the profitability of tearing down old structures and constructing anew. In this respect, also, the net effect of public housing construction may be less than the gross number of units built.

I am not aware of any attempts to measure the net effect of Canada's postwar public housing programs.

OTHER RENTAL SUBSIDY PROGRAMS

Governments have, over time, tried several approaches to assist low-income renters. Public housing projects are just one of these. Other programs have been developed that subsidize the private sector to provide low-rent accommodation. Some of these programs have been aimed at commercial builders and landlords, while others have been oriented to nonprofit groups and co-operatives. Still others provide subsidies to renters (e.g. shelter allowances) rather than to landlords.

The earliest postwar rental subsidy program actually began during the Second World War. In 1941, Wartime Housing Limited was created to provide accommodation for war workers in areas of housing shortage. This was not explicitly viewed as a low-income rental subsidy scheme, although it was geared to workers of modest means and provided decent housing at moderate rent. As the war wound down, this program merged with the Veterans' Rental Housing Program. About 50,000 rental dwellings were constructed under these two programs, mostly small, detached dwellings. In the late 1940s and early 1950s, most of this housing stock was sold off. The Canadian Council on Social Development (1977, 3) notes that the number of dwellings sold off (approximately 37,000 by 1954) greatly exceeded the 12,000 public housing units constructed between 1949 and 1964.

The 1944 National Housing Act provided several subsidies for builders of rental housing. One provision authorized joint loans for rental construction on generous terms. This was intended to promote construction of residential rental accommodation in general, not just low-rent housing. As with other NHA mortgages, these joint loans were replaced by loan insurance in 1954. Other provisions were specifically aimed at the production of low-rent housing. One authorized direct loans, again on generous terms, to "limited dividend" corporations that built and operated low-rent housing. Another guaranteed a minimum 2.5 per cent return to approved lenders on their investments in low-rent projects. Finally, a Rental Insurance Plan offered long-term, low-interest loans to builders of low-rent housing and guaranteed a total rent revenue that would yield a 2 per cent net return to the landlord. About 19,000 units were constructed under the Rental Insurance Plan before it was terminated in 1954.[20] The other two low-rent subsidy schemes continued until 1964.

Under them, about 107,000 rental units were constructed with direct, joint, or insured loans to private builders, not including the 28,000 units constructed by limited-dividend corporations.[21] All told, over 150,000 such rental units were built between 1946 and 1964, considerably more than the 12,000 public housing units constructed during that time.

With the 1964 NHA amendments, nonprofit corporations became eligible for loans to construct low-income rental housing. Through 1973, this provision was used to finance an additional 18,000 new rental units. In addition, CMHC insured mortgages for the construction of over 265,000 new dwellings. Further, after a brief interlude from 1965 to 1967 wherein CMHC withdrew from such activity, direct lending to other private-sector developers of low-rent housing was resumed. By 1973, 51,000 additional units had been constructed. Between 1964 and 1973, a total of 324,000 rental units were constructed under NHA.

The 1973 NHA amendments further extended assistance in the construction of low-rent housing. The amendments recognized "housing cooperatives" for the first time and treated them like nonprofit corporations. Further, co-ops and nonprofits became eligible for 100 per cent financing and for outright contributions of up to 10 per cent of construction costs. Between 1974 and 1981, CMHC made direct loans to nonprofit corporations for the construction of 46,000 new low-rent units and to co-operatives for just under 8,000. In addition, NHA-insured loans were made to co-ops and nonprofit corporations totalling 29,000 units between 1975 and 1981.

In 1975, the federal government introduced the Assisted Rental Program. The program at first provided an annual operating subsidy for each new rental dwelling. The subsidy was available for from 5 to 15 years to landlords constructing rental housing under a direct CMHC loan or an NHA-insured mortgage. Later, the subsidy took the form of an interest-free second mortgage for up to 10 years. About 125,000 units were constructed under this program before its termination in the late 1970s.

All the above programs encouraged construction of "private" rental accommodation. By private is meant both that the housing was not government-owned and that it consisted of private, i.e. not collective, dwellings. Various levels of government also provided subsidies for the construction of collective dwellings such as post-secondary student residences and old-age homes. These programs began effectively in the 1960s, and by 1970 more than 15,000 seniors and 70,000 students were housed under NHA-sponsored programs.[22]

In addition to new construction, the federal government subsidized the renovation of the existing rental housing stock. As was the case with owner-occupiers, a variety of programs were employed over the years. Very little is known about the net effects of these programs on housing stock production. Lithwick (1978b) looked at the net impact of the Assisted Rental Program (ARP). Using the Bank of Canada national econometric model, he attempted to predict the dollar volume of residential mortgage approvals in 1976 in the absence of ARP. The net contribution of ARP in 1976 was estimated to be about $99 million, or 4,600 net new rental dwellings. Given that there were 25,290 ARP unit approvals in 1976, this suggests that most of the ARP units would have been built in any case. However, Lithwick points out that the above method does not take account of the decline in the profitability of rental housing investment. In other words, the econometric model may have greatly overpredicted the number of rental housing units in 1976 without ARP. L.B. Smith (1981, 346) agrees that the net effect of ARP may have been large but attributes this to the presence of rent controls. He thinks that there would have been little incentive to invest in rental housing in 1976 in the absence of a heavy subsidy.

TAX SUBSIDIES TO LANDLORDS

Various provisions of the old and new Income Tax Acts gave subsidies to private landlords. One was the handling of depreciation expense. Almost any physical asset, be it a piece of machinery or a building, eventually wears out, even with normal maintenance. A depreciation expense is an allowance for this that effectively spreads out the purchase price of an asset over its useful life. In the calculation of income tax, some form of depreciation expense has always been allowed. Prior to 1954, the taxpayer determined the depreciation expense, subject to government approval.[23] Subsequently, capital cost allowance (CCA) was instituted to replace depreciation expense. The old and new acts defined classes of assets and a fixed CCA rate applicable to each. For example, wood frame buildings were in a class at 10 per cent CCA, while steel and concrete frame structures were in a class at 5 per cent. Further, the CCA rate was applied to a diminishing balance, making it the largest in the first year and declining to zero subsequently.

It is generally thought that CCA rates applicable to rental housing were too high in practice: i.e. buildings tended to have longer useful lives than that indicated by a 5 per cent or 10 per cent CCA.[24] In 1977, the CCA rate for wood frame buildings was reduced to 5 per cent to correct for this. Because CCA is based on the declining balance, a high rate overstates depreciation in the early life of an asset and understates it in later years. This shifts the temporal stream of income taxes paid by the landlord, reducing tax exposure early on and raising it later. This creates two advantages for the landlord. First, the landlord has a higher after-tax income in the early years and can use the additional funds to generate further investment income. Second, the landlord may currently be in a high tax bracket but anticipate being in a lower bracket later on, when the CCA is depleted. A high CCA rate thus permits landlords to defer income taxes until their marginal tax rates are lower.

Thus CCA formed an "accidental" subsidy to landlords in the sense that it did not represent an explicit housing policy. On at least one occasion, however, the federal government did use depreciation expense as a tool in promoting residential construction. This was the Double Depreciation provision introduced in 1947 that allowed landlords to use high depreciation rates, for up to 10 years, on rental housing projects approved by CMHC.

A landlord who constructs or purchases a building incurs a variety of "soft costs." These include mortgage insurance and guarantee fees, other costs of obtaining financing, mortgage and real estate tax costs during construction, and marketing and advertising fees. For the most part, these were initially treated as current expenses, to be subtracted from current revenues in calculating income for tax purposes. During the 1970s, the federal government shifted its stance and regarded these soft costs much the same as equipment purchases, i.e. amounts to be expensed (depreciated) over a usable life. If one adopts this view, allowing landlords to expense such items fully in the year that they occur reduces the landlord's tax liability in that year and raises it in future years, relative to what would occur with CCA. Thus the treatment of soft costs is a second "accidental" subsidy to landlords.

An important difference between the old and new acts was in the treatment of losses on rental property for tax purposes. Under the old act, a landlord could charge losses in the operation of rental property against any other income (e.g. profits on another venture, or salary income). This included rental losses arising because of CCA. Under the new act, most landlords could not claim losses against other income if the losses were created by CCA. Interestingly, this limitation did not apply to life insurance companies or corporations whose principal business was property rental. In effect, the new act made rental housing less attractive to small investors with other income that could otherwise have been "sheltered" by such rental losses.

In 1974, the federal government sought to promote rental construction by introducing the Multiple Unit Rental Building (MURB) scheme. Under this scheme, CMHC would certify a new building as a MURB, which entitled the owners to claim a full 5 per cent CCA, even if it created a rental loss. In other words, the MURB scheme essentially allowed small landlords to claim a rental loss against other forms of income, much as had been the case under the old act. The intent was to draw the small investor back into the rental housing market. The MURB program was discontinued in the late 1970s, then recontinued, only to be terminated in the early 1980s. Over its short life, MURB schemes totalling approximately 195,000 units were approved.[25] This formed almost one-third of the total starts of apartments and row houses during this period. Dowler (1983, 45) cautions against over-emphasizing the importance of the MURB scheme: "The MURB program's net impact on supply is much less than the total numer of certified units. This is due to three factors: (1) many certified MURBs never went ahead; (2) construction of MURBs often result in the demolition of existing affordable rental housing; and (3) a larger proportion of MURBs will revert to condominium tenure once the tax benefits run out."

It is also important to note that the Assisted Rental Program (ARP) coincided with the MURB program. Many MURB schemes were constructed with assistance from ARP. This makes it still more difficult to assess the impact of the MURB scheme itself on the overall supply of rental housing in Canada.

Before leaving the issue of depreciation allowance as a subsidy, we should mention "recapture." Under the new act, when an asset such as rental housing is sold, the landlord incurs a capital gain (or loss), which is the difference between the selling price and the original purchase price (or its market value at the end of 1971). Under the old and new acts, the landlord also incurs recapture (or a terminal loss, if negative) which is the difference between the lesser of either the original purchase price or the selling price and its undepreciated capital cost (i.e. the depreciated value of the building). Recapture was treated as normal income, while the capital gain either was ignored (the old act), or half taxed (the new act).[26] In effect, recapture brings any discrepancy between CCA and economic depreciation (i.e. as valued in the market) back into the landlord's income in the year of sale. Recapture arises because the CCA exceeded economic depreciation and thus represents a shifting of tax burden to later years, with the potential individual benefits as discussed above.

LAND SUBDIVISION

Between 1946 and 1981, Canada's urban population increased rapidly, increasing the number of households and the stock of dwellings. In part, this growth took the form of redevelopment of existing urban areas, i.e. existing low-density housing and other urban land uses being replaced by higher-density forms of housing. However, such redevelopment activity was only a small portion of total housing stock growth. Most of the housing stock change took place in the urban fringe. Vast suburbs were constructed on what had been agricultural or forested lands at the edges of existing cities.

In the early postwar period, there were relatively few controls on land owners who wanted to carve up large parcels of farm or forested land into suburban lots for detached housing, apartment buildings, or other such structures. However, there was increasing recognition of the fact that uncontrolled land subdivision could give rise to several physical, social, and economic problems. Urban sprawl, inefficiencies in provision of public services, and expensive remedial programs - such as replacement of septic tanks with municipal sewage systems - were among these problems.

The provinces gradually began to adopt land subdivision policies. In many cases, land owners could not subdivide land without government approval. Typically, broad discretionary powers were given to governments in approving subdivision application. Sometimes the bases for such discretionary powers were not even spelled out in legislation. For example, although Ontario's planning legislation dates back to 1946, it was not until the Planning Act of 1983 that the government explicitly stated its bases. Even then, the provincial interest was put quite broadly: protection of the natural environment and features of significant natural, architectural, historical, or archaeological interest; conservation of energy; provision of major communication, servicing and transportation facilities; equitable distribution of social facilities; co-ordination of planning activities; resolution of planning conflicts; health and safety of the population; the financial and economic well-being of the province. In addition, the province stated that it would look at the following specific features of any subdivision plan: whether it is premature or in the public interest; whether it conforms to the official plan and adjacent plans of subdivision; the suitability of the land for the purposes proposed; the dimensions and shape of the lots; proposed restrictions on the land, buildings and structures proposed to be erected; conservation of natural resources and flood control; the adequacy of highways, utilities, municipal services, and school sites; the land to be conveyed for public purposes. In effect, such broadly stated interests largely gave the provinces carte blanche to pursue whatever policies they desired.

It is not clear if the provinces realized that land subdivision policies would affect aggregate growth of the housing stock. However, some simple economic principles suggest unmistakeable effects. Subdivision control restricts the supply of land available for suburban expansion. Supply restrictions generally raise prices. Builders, faced with more costly land, put up housing that is more land-intensive. One result was higher-density housing than would have occurred in the absence of land subdivision controls. In addition, as housing became more expensive, the pace of household formation was slowed; hence a smaller increase in the housing stock than might have been expected without controls. Thus a smaller housing stock with greater emphasis on apartments and row-housing (rather than detached dwellings) may have been a consequence of subdivision controls.

However, the above argument is speculative. It assumes that such controls drove up the price of developable (i.e. subdivided) land. It is not clear, though, just how much these controls contributed to the rising price of suburban land over the postwar period. If the supply of approved subdivision land was always large enough to satisfy demand, there would have been little upward pressure on land prices. It is possible, in other words, that land subdivision controls did not substantially affect land prices and hence the growth and character of the housing stock. Indeed, some local governments approved substantially more subdvision than was needed, given the subsequent demand for subdivided lots.

Critics of subdivision regulation often make international comparisons in arguing that Canadian controls forced up the price of land. In the United States, for example, land subdivision control was less prevalent and less severe than in Canada. There were marked differences in land prices, lot sizes, and mixes of housing types in comparable locations between these two countries. In general, the Canadian location was less likely to have a detached dwelling on a larger lot and was likely to be more expensive. The extent to which such differences are attributable to subdivision control is unclear, though, given the variety of other factors that might also account for such differences.

More important may have been the impact of land subdivision control on the temporal sequencing of housing costs. In uncontrolled suburban development, there was a tendency for expensive public services to be installed long after the initial residential construction. For example, it was not uncommon for an uncontrolled suburban area to be initially developed without street lighting, sidewalks, municipal sanitary or storm sewers, or garbage collection. Over the succeeding years, these were often added incrementally, and sometimes not at all. Typically, these costs were passed on to residents at that time, either as a lot levy specific to the beneficiary or spread over all local taxpayers. With land subdivision control, many municipalities came to insist that the developer arrange for such amenities prior to, or in conjunction with, residential construction. The cost of these "minimum servicing standards" was put onto the consumer from the outset.

In addition, some local governments insisted that the entire cost of development be paid upfront prior to subdivision approval. Others allowed servicing to be financed by a bond and paid for by lot levies in the approved subdivision. The latter scheme spreads the cost of servicing over a number of years. The former forces the consumer to pay these costs upfront or spread them out through mortgage repayment. In either case, the cost of housing under subdivision control reflected these servicing standards. Without subdivision control, the consumer paid a lower initial price and incurred additional charges only if and when the amenities were installed.

This "forward shifting" of costs to the consumer may well have reduced the affordability of suburban housing. The primary market for such housing was young families, the kind of household with limited current income but good prospects for future growth. Such households might be better off to defer certain costs, such as the upgrading of amenities, until later, when their incomes were higher. However, subdivision control had the effect of forcing them to purchase a given package of servicing standards at the outset.

Opinions differ on the impact of servicing standards. Some planners argue that it is more efficient to provide a full package of amenities at the outset than to upgrade.[27] Others argue that the real problem is that the servicing standards are simply too luxurious, that households do not want the amenities they are forced to purchase. In this sense, land subdivision control needlessly made suburban housing less affordable.

Metropolitan Toronto provides an interesting case study. In 1954, the province stopped approving subdivision plans that were based on septic tank systems.[28] Subdivisions were approved only if they could be connected to a municipal sewage system. This severely constrained development on the northern fringe of Toronto, where topography and distance to Lake Ontario (the "sink" for all treated sewage in the area) made extensions to the existing sewage collection system difficult or expensive. A household wanting to purchase a new suburban home usually found itself restricted to one of the approved subdivisions. The dwelling typically would have municipal water and sewage connections; abutting public streets would be paved, lit, and be regularly cleared of snow. The alternative was to purchase a larger lot in the countryside. Typically this was a farm or bush lot no smaller than about four hectares (ten acres).[29] For the most part, these lots were limited to well water and a septic tank. Local roads were often unpaved and unlit, and snow clearance was irregular at best. In addition, there were relatively few of these lots. For many households, there was no choice. The restricted availability of exurban lots and their poor servicing made them unviable. However, approved subdivisions tended to be quite similar in amenities and design, since all were approved by the same provincial agency. Hence there was little variety in housing alternatives.

Over the postwar period, land subdivision control became more sophisticated. Governments increasingly scrutinized subdivision applications and entered into long negotiations with developers. The result was delay in the subdivision approval process. In a study of Mississauga, Ontario, Proudfoot (1981, 286) found evidence of this: "Ten to fifteen years ago, it took an average of a little over two years to process a plan of subdivision from submission to registration; in recent years it has been taking well over three years."

Whatever the benefits of such regulation, there were two distinct kinds of incremental costs. One was the added cost of compliance. Increasingly, a developer had to present additional technical information (e.g. an environmental impact statement) and to understand complex legal contracts. To do so, the developer had to rely on a technical staff or hired consultants, and this was not inexpensive. A second kind of cost arose from the delay associated with regulatory compliance. Such delays typically imposed extra costs on the developer, including the interest cost of funds invested in the project while awaiting approval. In this respect, the delay was reflected in higher costs of bringing subdivided land to market, and hence higher downstream costs for housing built on that land.

How large an effect did such incremental costs in subdivision approval have on the price of, and subsequently on the demand for, housing? This is unclear because there is little evidence on the magnitude of such costs. In one of the few studies of this type, Peiser (1981) analysed subdivision lot prices in Dallas and Houston, Texas. In Houston, there were fewer regulations governing land subdivision and shorter delays. It was found that about $1,000 of the average lot price differential could not be explained by other factors (such as differences in the method of financing utility improvements) and therefore presumably reflected the differential effect of subdivision regulation. Given an average lot price in Houston of $13,850 at the time, the incremental cost of the subdivision approval scheme in Dallas was under 10 per cent. Of course, this cost may well differ for other jurisdictions with different land subdivision controls.

ZONING AND RELATED REGULATORY CONTROLS

Other federal, provincial, and local regulations also helped shape postwar housing stock change. At the provincial and municipal levels, these included zoning bylaws and building codes. At the federal level, the role of CMHC standard setting is noteworthy. Over the postwar period, these regulations became increasingly sophisticated.

As with land subdivision control, these other regulations were presumably introduced because they created benefits for society. At the same time, any regulation imposes certain costs, either to individiuals or to society as a whole. Many of these regulations may have been worthwhile in the sense of benefits outweighing costs. However, there is little in the way of rigorous policy analyses to support such a belief.

What was the effect of such regulation on the cost of building and, hence, on the amount of residential construction? As in the case of land subdivision above, these regulations imposed costs in terms of added delay and compliance. How large were these costs? How did they change over the postwar period? To what extent was aggregate demand for housing reduced as a consequence of the attendant higher prices for new housing? Unfortunately, there is little in the way of direct evidence by which these questions might be answered.[30]

Other effects of zoning and other regulatory controls tended to be specific to the regulation concerned. Let us now consider some in more detail.

Zoning Bylaws

Every Canadian province made it possible for local governments to pass "zoning" bylaws. The exact legislation varied from one province to the next. However, in general, local governments were permitted to control the pattern of land use and the form of construction. In Ontario's Planning Act of 1983, for example, zoning bylaws could be passed to prohibit certain uses of land, or the erection or buildings, in particular areas; to regulate the type of construction and the height, bulk, location, size, floor area, spacing, character, and use of buildings; to regulate the minimum frontage, depth, and area of a parcel of land, the proportion of the lot area that a building or structure may occupy, and the density of development; to regulate the minimum elevation of doors, windows, and other openings in buildings; and to require the provision and maintenance of loading or parking facilities.[31]

What effects did zoning bylaws have on the aggregate housing stock and its composition? As the above description of the Ontario legislation attests, zoning bylaws served a number of purposes. For some of these, the effects of zoning bylaws on the housing stock may have been negligible; for others, not.

Commonly, one purpose of a zoning bylaw is to improve the locational pattern of land uses. A bylaw may, for example, prohibit noxious land uses in residential areas, encourage a mix of residential and nonresidential land uses so as to reduce inefficient trip-making, or prevent development in floodplains or other unsuitable areas, provide for parkland and other community spaces, or encourage integration of architectural and site design within a local setting. A developer wishing to construct an apartment building or a subdivision of row-housing, for example, might find that a zoning bylaw restricts such housing forms to particular locations. However, such a bylaw need not have an effect on the aggregate housing stock, provided there is a good selection of comparable alternative locations. In other words, the zoning bylaw may affect the location of a particular form of housing, but not the total amount built. Of course, this conclusion depends on the availability of "a reasonable selection of comparable alternative locations."[32] A zoning bylaw that severely restricts the possible sites for a particular form of housing may well help shape the size and composition of the housing stock.

However, in some cases, zoning bylaws are also used to exclude a land use from the local area altogether. For example, a bylaw might set minimum lot sizes or floor areas, the effect of which is to restrict the supply of low-income housing. Such zoning can affect the aggregate growth and composition of the housing stock. At the least, it alters the mix of dwellings, resulting in more large, detached dwellings and fewer small, attached, or apartment dwellings. In addition, some individuals, for whom a smaller dwelling would be affordable, are unable to afford the available larger units. Rather than forming separate households, these individuals remain as lodgers or family members in existing households, thus reducing the aggregate demand for dwellings.

Although exclusionary zoning may have affected the size and composition of local housing stocks, its empirical significance is not evident. In general, provincial governments have not looked favourably on exclusionary zoning, especially when used to "zone out" low-income groups. However, a zoning bylaw can be quite complex, and it is difficult in practice to identify and prevent exclusionary or near-exclusionary provisions. Consider the Ontario legislation as an example. In specifying "the minimum elevation of doors, windows, and other openings in buildings," it may be possible to exclude dwellings such as basement apartments that would normally provide low-income housing.

In postwar suburbs, one particular form of exclusionary zoning was widely practised. This was a zoning provision that prevented owners of single detached dwellings from adding basement or upstairs flats (thus converting their structure into a duplex). Some owners ignored these rules and illegally added a flat in any case. However, for the most part, these zoning provisions restricted the supply of flats and thereby presumably affected both the size and composition of the aggregate housing stock.

Another purpose of zoning bylaws is to preserve existing neighbourhoods or other local features. In inner city areas, for example, there may be substantial pressures to demolish older dwellings and construct new housing of a type more profitable, given current market conditions. In the late 1960s and early 1970s, there was pressure to develop high-rise apartment buildings in older, low-rise neighbourhoods. Later, there was more pressure for deconversion of large dwellings from flats back to single detached units (as part of "whitepainting" or "gentrification"). In both cases, concern was expressed for the preservation of the existing neighbourhood and housing stock, and sometimes zoning bylaws were used to facilitate such trends. However, as a result, the cost of new housing construction may have been increased, and some construction forgone.

Building Codes

Laws governing construction and material standards originated in Canada at the municipal level. The municipalities had been expected to handle the problems that arose from the density of urban development: the spread of fire, the collapse of large buildings, poor sanitation, and so on. A building code for new construction was necessary to prevent many of these problems.

Of course, a municipality could not simply enact a building code bylaw. First, the province had to pass enabling legislation. Prior to the 1970s, provincial legislation was largely permissive. The municipalities were, in general, left to determine the exact form of their bylaw, the aspects of building to be regulated, and the minimum standards to be imposed. This was no small undertaking. Such a bylaw could easily run to several hundred pages of technical points. To create such a bylaw and to enforce it presumed a trained and experienced technical staff, something that not every municipality could afford. It has been estimated that, in the early 1950s, there were about 4,000 municipalities in Canada, many of which had their own building code. Undoubtedly, there was a lot of borrowing: city A adopting the building code used in city B.

However, as each municipality amended its code over time to suit its own needs and problems, considerable diversity of standards arose. Sharp regional differences were evident in the housing stock right across Canada. Housing materials and construction techniques that were quite acceptable in the Maritimes might not be legal in parts of Ontario, and vice versa. Further, building codes were used to promote several objectives including protecting against building failure or collapse, improving fire resistance, improving safety standards and hygiene, and promoting a variety of aesthetic and urban design objectives.

The National Research Council first produced a model national building code in 1941. A second, more comprehensive edition was completed in 1953, and revisions have occurred regularly since then. After 1953, federal and provincial governments began to pressure the municipalities to adopt the national code.[33] Many municipalities did so, although they did not necessarily update their bylaws with each subsequent revision of the national code. In the 1970s, the provinces began to take over building code regulations themselves. In Ontario, for example, the Building Code Act of 1974 specified a single province-wide code based on the national model. It superseded all municipal building code bylaws. For the first time, building construction was under the same set of regulations throughout the province. To summarize to this point, the postwar period saw a shift from a municipally based code system with substantial local variations, to a provincially based system broadly similar to the national building code.

Little empirical evidence is available of the effects of building codes on development of the postwar housing stock. However, casual observation suggests that they may have been substantial. The sheer length and detail of a typical building code posed two additional types of costs for builders. First were the compliance and delay costs common to all forms of regulatory constraint. Builders and tradesmen needed better technical knowledge; there was the red tape and delay associated with obtaining a building permit and having the required inspections done; there were the costs of ripping out and redoing work deemed to be substandard; and there were inefficiencies for builders because the standards varied from one locality to the next. Second, there were costs involved in building at code standards, compared to substandard practices. Undoubtedly, these incremental costs meant that housing was more expensive. How much more expensive, and consequently how large an effect they had on household formation and on the aggregate growth of the housing stock are unclear.[34]

At the same time, opponents argue that some of the code provisions are unnecessary. In certain cases, they see the standards as being too luxurious, arguing that a lower standard would suffice. This is akin to the argument about subdivision servicing standards raised above. An example of this was the requirement, in some local building codes, of brick exterior walls. Early experiences with large fires led some cities to insist on brick walls. At the time, this was the most fire-resistant sheathing material available. However, advances in building technology, including aluminum siding, created new possibilities for siding materials. By ignoring these, local building codes in effect forced builders to use a standard that was more expensive than necessary.

Another effect of building codes is that they can discourage certain types of residential construction. This is particularly true of conversions. Suppose, for example, that an owner wishes to convert an older, single detached home into a duplex. Suppose further that, being older, the building no longer meets current building code requirements. If the structure were to continue as a single detached dwelling, no upgrading would be required unless it were to fall short of some property maintenance regulation. If the owner wants to proceed with the conversion, however, some upgrading of the building to current standards would likely be required. This upgrading would be a cost over and above the cost of the conversion itself. Although some might convert illegally, the net result is that some landlords would be discouraged from converting.

Standard Setting by CMHC

This chapter has already considered the extensive federal direct lending, loan insurance, and grant operations under NHA. As part of these operations, the federal government established sets of housing standards that eligible housing would have to satisfy. Consider, for example, a builder who wants to put up a new detached dwelling that would be eligible for an NHA-insured mortgage. Central Mortgage and Housing Corporation (1955) is a 129-page booklet outlining minimum standards that have to be satisfied. Among the standards for a detached dwelling are the following. A dwelling could not be built on a lot less than 40 feet wide or with fewer than 4,000 square feet (7,500 if a septic tank were present, and 15,000 if both a septic tank and a well were to be present). The dwelling could not cover more than 33 per cent of the ground area of an inside lot, or 40 per cent of a corner lot. Accessory buildings could not cover more than an additional 10 per cent, nor exceed one storey in height. The minimum distance from a building to the street line was 12 feet; the minimum depth of rear yard was 25 feet; and minimum sideyard width was 4 feet plus 2 feet for each additional storey above the first. Driveways could not be less than 8 feet wide and must be entirely on the lot served. These requirements dictated that detached dwellings be constructed at a low density, with extensive front and rear yards. Further, dwellings could not be NHA-financed if built on "unusual" lots: e.g. lots with too small a street frontage or with no frontage at all. To the extent that housing was not likely to be built without NHA financing, these standards affected the kind and number of dwellings built during the postwar period.

At the same time, the above standards were only a small part of the total set of CMHC requirements. Essentially, CMHC regulations also specified a simplified building code. There were, for example, minimum standards governing foundation walls, structural members, drainage, masonry construction and veneers, wood frame construction, roofing, insulation, and sheathing, fireplaces and chimneys, doors and windows, interior and exterior trim, fire protection, and plumbing and electrical work. In addition, there were a variety of minimum standards in regard to the internal layout of the dwelling. These include minimum size, floor area, and ceiling height for each room; minimum configuration of hallways (including a front entrance hall), with minimum widths; and minimum number of closets, with internal dimensions and locations. At least one bathroom was required. It had to be accessed from a hall linking the bedrooms. Other bathrooms could be off the bedrooms. No habitable room could be constructed in a "cellar" (i.e. a basement which than one-half below grade). Each "habitable" room had to have one or more windows. The total window area had to be at least 10 per cent of the room's floor area. Bathrooms, halls, and vestibules are not counted as habitable in this respect.

These latter standards contributed, I believe, to a standardization of housing constructed in the postwar period. In choosing a new dwelling, households were faced with either an NHA-financed unit built according to these standards or a conventionally financed dwelling with perhaps different characteristics but at less favourable financial terms. In effect, the subsidies implicit in NHA programs helped to tip consumer demand toward housing based on those standards.

CONCLUSIONS

This chapter has documented the extensive role of governments in shaping Canada's emerging postwar housing stock. Although public housing itself never formed more than about 2 per cent of the housing stock, overall the influence of government was quite pervasive. Successive federal governments since 1936 have been committed to restructuring the mortgage market so as to induce "easy money" for new housing construction. In addition, a variety of federal, provincial, and local programs over the years have offered grants and favourable loans for renovation and upgrading of the existing housing stock, especially for owner-occupiers. Further, a variety of tax expenditure programs have effectively reduced the cost of home ownership or increased the profitability of (or reduced the risk of investment in) rental housing. Finally, through various regulatory requirements, federal, provincial, and local governments have affected both the price of new housing and its characteristics, thereby affecting the growth and composition of the housing stock.

What has been the net impact of government on the postwar housing stock? What might the housing stock have looked like in the absence of a particular housing program or policy? In general, it is not possible to answer such broad questions from the fragmentary evidence described in this chapter. It is clear that a large proportion of dwellings in Canada were financed through NHA, that almost all were constructed under a variety of federal, provincial, and local regulations, and that all owner-occupiers benefited from certain income tax provisions. However, the central question is: How many fewer dwellings, and what other mix of dwellings, might have been constructed in the absence of such public-sector involvement? Presumably, some amount of housing would have been built anyway, in view of the population changes and the growth in incomes noted in earlier chapters. Just what might have been built under another set of government housing policies is unclear.

CHAPTER TEN

Conclusions

Between 1945 and 1981, Canada experienced some remarkable changes in its housing stock and patterns of household formation. While the population doubled, the number of households trebled. If measure of change in the overall housing stock – including secondary dwellings, vacation homes, unoccupied dwellings, and so on – were available, I suspect it would show even more rapid growth. Further, there is evidence that postwar homes became larger (compared to the size of the household occupying them) and better appointed.

Setting up a separate household, or having a larger home or second dwelling, is typically quite expensive. In addition, one's choice of living arrangement is partly an affair of the heart. In many cases, one lives with a spouse, children, parents, friend, lover, or partner out of a sense of desire, need, obligation, or duty. For most people, these are not decisions made lightly. Changes in patterns of household formation and housing demand thus must reflect substantial changes in affordability or social structure.

The objective of this book is to describe, explain, and link these factors. This is a difficult task. A rapidly changing social and economic climate makes it hard to assess the importance of any one particular determinant. The research literature is strewn with examples of studies that consider just one important determinant. These have been largely wasted efforts. What good is it, for example, to look at a relationship between living arrangement and income over time, without carefully taking account of changing demographic structure? Looking at one factor in isolation easily leads to mistaken interpretation. For example, it is conventionally thought that the propensity to be a home owner rises with income; however, the postwar period includes sustained periods of real income growth where the likelihood of ownership in fact declined. This does not mean that income was not important or that its effect was negative. Rather, it illustrates the importance of looking at the interaction of several demographic and economic factors.

A central theme of this book is that no one single argument adequately reflects the extent of what happened. Several different kinds of demographic trends were important, as were several economic changes. The research problem is to find ways of identifying their separate contributions. Unfortunately, as in much social research, we cannot look at historical experience as a kind of laboratory experiment wherein all the explanatory factors but one are tightly controlled. It means that we have to be careful in distinguishing which factors underlie a particular comparison.

METHODS OF ANALYSIS

Unravelling the effects of several factors is a kind of detective work. As with good detective work, precision is paramount. In one part of this book, for example, I refer to tenure patterns among four-person urban families living alone and headed by a spouse-present male aged 35 to 54. Such fine disaggregations do not make for exciting literature or sweeping overviews. However, they can be insightful. Disaggregation permits us to control for other factors and look in this case at the effect of income itself on tenure. Income was found to be important in shaping tenure, but only for certain demographic groups. This kind of careful decomposition of existing data is necessary if we are to understand better what happened. Concern for precision through disaggregation underlies much of this text. The methods of analysis in chapters three, four, five, six, and eight in particular evidence this approach.

For the most part, this book looks at household formation and housing stock change using demand analysis. However, choices of living arrangement and dwelling are also conditioned by what is available: i.e. by supply factors. This demand analysis presumes that housing markets are in an equilibrium that is manifested in the prices of housing alternatives. Therefore the response of consumers to changing prices is, at least in part, a response to changing supply factors. Although this book takes price changes as given, we should remember that these may well reflect changes in the determinants of supply.

The one exception to the general exclusion of supply factors comes in chapter nine. There attention is paid to the role of governments in shaping housing stock change. Almost all postwar housing construction in Canada took place under the provisions of various government regulations, and much new and existing housing received some kind of financial assistance. The potential effects of such activities on household formation and housing demand are too large to ignore or treat indirectly.

In demand analysis, consumers' choices of living arrangement and dwelling are seen to be determined by their tastes, income, the prices of alternative residential forms, and the prices of other goods and services. Tastes in turn are related to various demographic characteristics of the consumer. Thus demographic and economic (prices and incomes) perspectives on household formation and housing demand arise naturally.

This raises a quandary regarding the definition of a consuming unit. Some housing demand studies look at the choices of individuals, others at the choices of families (and/or nonfamily individuals), and still others at the choices of households. None is inherently satisfactory. In the terminology of chapter two, the individual is a rigid unit of analysis. However, most individuals form part of a family unit and pick living arrangements that reflect this. It would be silly, for example, to attempt to measure the impact of affordability on the living arrangement choices of young children. In this sense, the individual is not a relevant unit of analysis. The family or the household provides a more relevant unit of analysis. However, neither is rigid; a family or household is by definition a co-resident group. Taking ito be the consuming unit ignores the fact that some family or household members may respond to changed conditions by moving out, hence possibly altering the number, size, and composition of households. Many studies of housing demand look at the dwelling choices made by households based on their incomes and tastes and the prices of housing alternatives but ignore the effects of housing prices on the formation of households themselves. One objective of this book is to emphasize that link.

There have been few attempts to date to look jointly at household formation and the demand for housing. As with this book, those attempts look at household formation and housing demand sequentially. They have not modelled the behaviour of individuals seeking out living arrangements and living quarters as simultaneous decisions. Rather, as with this book, they offer certain models to describe or explain household formation, followed by additional models to describe or explain the housing demands of these households. An important area for future research will be the formulation of simultaneous models.

POSTWAR HOUSEHOLD FORMATION

Prior to moving into this area, my research focus was on models of urban economic growth. In particular, I was interested in the relationship between economic growth and population change. It had become evident to me that changes in age structure – through their effects on household formation, and hence on the demand for different kinds of housing –

were as important as overall population growth in determining urban size and form. Influential in my early thinking on this subject were the empirical works of David (1962), Beresford and Rivlin (1966), Gottlieb (1976), and Hickman (1974).

Newcomers to the area quickly become familiar with the use of headship rates. Virtually all long-range forecasts of household formation are made by applying headship rate estimates to projections of population by age and sex (and possibly marital or family status). The widespread use of headship rates in projections gives the impression that these rates are stable (hence predictable). It is easy to fall into a trap wherein headship rates, like mortality, fertility, or migration rates, are thought to change only slowly over time. In this trap, household formation is seen as largely driven by demographic forces (births, marriages, divorces, deaths, and widowhood). However, this view does not square with the facts. While it is true that headship rates are stable (in fact they are virtually 100 per cent for most husband-wife families), they have been considerably more variable (and lower) for other kinds of consumers. Hickman (1974) presents compelling evidence of the variation in headship rates over much of this century and shows that this was closely linked to economic conditions. His work suggests the importance of economic factors in household formation. An analysis of the relative effects of demographic and economic factors in shaping household formation thus became a focal point for this book.

A review of the research literature on household formation yields a bugbear in the term privacy. Among those using the term are Beresford and Rivlin (1966). They argue that the postwar increase in families living alone cannot be explained simply in terms of rising prosperity and demographic shifts. In their view (p. 247): "The typical modern American apparently puts a high value on having a separate dwelling unit, into which he can retreat with his wife, if he has one, and his minor children but no one else, and close the door. He is reluctant to share a dwelling with relatives outside his nuclear family or to live as a roomer or boarder in the household of a non-relative. Since World War II, Americans have expressed these preferences by using part of their rising income to buy privacy. At all age levels, individuals and nuclear families have succeeded in obtaining not only more housing and better housing but housing separate from other people."

The tautological nature of this "increased desire for privacy" argument is bothersome. That a rising proportion of persons live alone is often taken as evidence that people desire more privacy. Possibly it does reflect a change in tastes. However, it could also reflect other factors, such as higher income. My own inclination is to look for other such arguments the validity of which can in some sense be tested. I suspect that changes in patterns of household formation can be adequately explained without invoking an argument about changing tastes.

What then brought about the postwar rise of the nuclear family living alone? Rising prosperity was important. However, it is not correct to say that husband-wife families used their growing real incomes to purchase privacy. Evidence suggests that the taking in of a lodger had little to do with income. Among both rich and poor families, the typical lodger was another relative, be it a grandmother, nephew, or brother. The taking in of lodgers as a commercial enterprise was not widespread. The rise of the nuclear family living alone thus amounted largely to a splitting off of relatives. It should not be surprising then that it was little related to rising prosperity among families.

Of more importance was the increasing incidence of working wives. For a variety of reasons, women began entering the paid work-force in ever greater numbers over the postwar period. On the one hand, this contributed to the rise in family incomes. On the other hand, it reduced the ability of the family to take in a lodger by cutting into the time available for housekeeping. In this view, the postwar rise in income was concurrent with, but did not necessarily cause, the decline in lodging. Rather, increased labour force participation among wives contributed to both.

Also important was the growth in income among nonfamily individuals. Evidence shows that their choice of living arrangement was quite sensitive to income. In other words, as their incomes rose, nonfamily individuals increasingly split off from families to set up their own households. It was nonfamily individuals, not families, who used rising incomes to purchase privacy. This growth in income was partly achieved by public policies - including Old Age Security, the Guaranteed Income Supplement, and the Canada/Quebec Pension Plans - designed to support the elderly poor.

The fly in this ointment is that it does not seem to explain pre-war household formation well. Through earlier parts of the twentieth century, up to the Depression, there were also some substantial gains in incomes. However, there was nothing like the explosion of household formation that occurred after 1945. We have already noted that, after 1951, there was a substantial change in the character of the one-person household. It became primarily an urban phenomenon, and one that was not tied to occupational structure. Why wasn't there evidence of this shift earlier in the century, when real incomes had also risen?

Several answers come to mind. One is the argument of Michael et al (1980) that the relationship between income and living arrangement is S-shaped. Central to this argument is the notion of a threshold income required for someone to live alone. Below that income, individuals must lodge or otherwise share accommodation. Above the threshold, they find living alone affordable. Over time, the threshold shifts in response to changes in the price of housing, the prices of other goods, the prices and availability of housing substitutes, and consumer preferences. However, with continued vigorous growth in real incomes, a point is eventually reached where living alone becomes affordable to a large number of nonfamily persons. There is evidence of an S-shaped (logit model) relationship between income and living alone, and by the 1970s many nonfamily individuals were on the sharply rising portion of that curve. Michael et al surmise that up to the 1940s most U.S. nonfamily individuals had incomes below the threshold, i.e. were on the flat lower part of the S-curve. Possibly the same is true of Canada. However, Michael et al present no evidence to support this conjecture, and I am not aware of anyone else who has.

Growing prosperity, although important, was not the only factor. Before the postwar period, many nonfamily individuals who could afford to live alone did not. As important as changes in income have been changes in home-making technology. Maintaining a home in a state of cleanliness, good operating condition, and repair has always been onerous. For younger nonfamily individuals, such maintenance activities compete with other valuable uses of time, i.e. work, leisure, or social activities. For the elderly, there may also be physical constraints on their abilities to maintain a dwelling. Over the postwar period, there were some remarkable changes in the range and prices of consumer goods, notably in the areas of cooking, housecleaning, and laundry. In terms of cooking facilities, one should note the development of dependable, precise electric and gas stoves, self-cleaning ovens, slow cookers, microwave and convection ovens, pop-up toasters, electric frying pans, a variety of timer-controlled cooking devices, and other power kitchen tools. Also important was the proliferation of "fast food" alternatives to lengthy home food preparation. In terms of housecleaning, vacuum cleaners, electric brooms and carpet cleaners, a variety of dust-removing and cleaning solvents, and more durable and stain-repellant furniture fabrics and carpets should be noted. Laundry chores were eased somewhat by innovations in electric clothes washers and dryers, the diffusion of coin-operated laundromats, electric irons, and no-iron fabrics. These lists are probably incomplete, but they illustrate just how much home-making technology eased the chores of living alone. It has not been possible, however, to estimate the impact of changing home-making technology on the decision to live alone. In part, it is difficult to quantify just how extensive these changes were. In part, the changes were so pervasive that it is unclear how their separate effects might be empirically separated. However, the relatively rapid postwar growth in living alone, compared to earlier times when incomes may have also been increasing substantially, is consistent with the argument that postwar improvements in home-making technology were important.

Another facet of this argument is the changing provision of in-home services. Earlier in the book, we noted that the majority of people living alone were elderly. Many of these elderly were able to live alone only because of the provision of certain in-home services. Included here are low-cost or free in-home nursing, "meals on wheels," and handicapped transportation programs. Before these programs were introduced, some of the elderly would have had to live in an institutional facility, a child's home, or another place where these services or substitutes could be provided. In effect, these programs helped to make living alone more feasible or inexpensive for a certain group of nonfamily individuals.

Another important contributing factor was the rise of assisted housing. Much assisted rental housing was built for the elderly in the 1960s and 1970s. The units were typically quite small (bachelor or one-bedroom), hence easy to maintain. For the most part, they were let on a rent-geared-to-income basis that made them affordable, even considering the tenants' relatively low incomes. Also, some of this senior citizen housing was provided with certain social and medical support services and in-house facilities that helped to make living alone more feasible or attractive. Also important among the elderly was the introduction of Old Age Security, the Guaranteed Income Supplement, the Canada/Quebec Pension Plan, and free or subsidized medical/hospital/drug insurance, not to mention the spread of private pension and disability insurance programs. Full indexing of public pensions in the 1970s added further to the income security of the elderly. These programs helped to make separate housing more affordable for the elderly, especially those who were not in assisted housing.

The rising incidence of living alone was also abetted by changes in the characteristics of nonfamily individuals. There are four basic groups of nonfamily individuals: young single adults, older bachelors (men and women), divorced persons, and the widowed.

One important life-style change is evidenced in nuptiality trends. Through the first half of the twentieth century, there is evidence of a declining incidence of bachelorhood at older ages. For a variety of reasons, marriage became more and more popular. This trend continued through the end of the postwar marriage rush. Then, in the 1960s, there began a switch away from marriage among young adults. In part, this was a postponement of marriage; in part, it also reflected a rising number of people who would never marry. In chapter eight, it is argued that these new singles differed from their predecessors. They were more upwardly mobile, had better incomes, and could better afford to live alone.

Another group whose characteristics changed were the widowed. For the most part, this group consists of older women. Through the postwar period, it was made up of women who had borne their children in the first three decades of the century. With fertility declining, older widows in the 1970s on average had had fewer children than their predecessors of the 1950s. It has been argued that a widow is much more likely to live with one of her grown-up children, the more children she has had. Therefore the rising incidence of living alone among widows in part reflects the decline in fertility that occurred from 1900 through the 1930s. It also reflects changes in geographic mobility. The postwar period was one of high rates of internal migration among young adults. In many cases, they left their parents behind when they moved. With widowhood, many parents were left with a choice between living alone in their present town or moving to a new town to live with a child's family. A rising incidence of living alone may simply reflect, in part, the parent's desire not to leave the old town.

POSTWAR HOUSING DEMAND

By definition, since every occupied dwelling contains exactly one household, the number of occupied dwellings kept pace with the formation of households. However, the total stock of dwellings in Canada also includes second homes, vacation dwellings, and other unoccupied units. There is no reliable measure of postwar change in the total housing stock; however, it is likely much greater than the net formation of households. The remaining discussion is, however, largely restricted to the stock of occupied dwellings.

Change in the housing stock can be viewed quantitatively: in terms of the number of dwelling units added. It can also be viewed qualitatively: in terms of the quality or characteristics of a typical dwelling. By several criteria, the quality of the occupied stock improved over the postwar period. Chapter seven notes some of the improvements, especially in rural housing. Electrification became universal, as did central heating, running water, and bath and toilet facilities. New building materials (e.g. plywood, fibreglass insulation, and vinyl and aluminum sidings) and the greater use of prefabricated materials (e.g. roof trusses, kitchen cabinets, and window frames) also contributed to dwelling quality (while keeping costs down).

Until the 1970s, the average size of a dwelling remained roughly constant at about 5.3 rooms, then it increased to 5.7 rooms by 1981. However, over the postwar period, the average number of persons per dwelling fell. Thus, the average number of persons per room, a commonly used measure of crowding, in fact declined. In this sense, as well, housing consumption increased. Households consumed more and more rooms relative to the number of persons living there.

Counts of dwellings, their numbers of rooms, and the presence of basic facilities (such as running water) are available from census sources. There are, however, few other housing quality data available on a time series basis. For example, we know little about changes in floor area or in typical room sizes. There can be little doubt, though, that the housing stock improved in several important, if not readily measurable, respects. For example, improvements in building technology and materials, and the introduction of building codes, undoubtedly have helped to ensure a housing stock that is more resistant to deterioration, collapse, and fire.

At the same time, other changes represented unclear or negative contributions to dwelling quality. As mentioned in chapter seven, quality change is partly in the eye of the beholder. In my view, the shrinking of ceiling heights to a standard of 8 feet (2.4 metres) represents a decline in quality, as does the replacement of wetwall by drywall construction and the abandonment of brick masonry walls. Not wishing to bemoan the passing of outmoded construction styles, I might argue that consumers made their choices; that they did not feel, for example, that the hollow sound or jointing problems of drywall were sufficiently bad to offset its cost savings. However, to the extent that wetwall was preferred to drywall construction (cost aside), its replacement does represent a reduction in dwelling quality.

While recognizing that any overall assessment of quality change is ultimately subjective, there have undoubtedly been some substantial quality improvements. What brought these about? In part, they reflect growing prosperity. Rising incomes permitted households to purchase larger and better accommodation. They have also permitted more affluent middle-aged families to support directly the separate living arrangements of grown-up offspring and of parents. This same prosperity also enabled Canadians to impose more restrictive building codes and standards. These higher minimum standards have been applied to all housing, not just for the wealthy who can best afford it. At the same time, the need to provide a high quality of basic housing services to low-income households has been addressed through various subsidy programs.

As important as the growth in nominal incomes was the relatively sluggish rise in the price of a standard unit of housing services. Available rent indices show a particularly low rate of increase over the postwar period. Price indices for owner-occupancy based on the cashflow cost of ownership rose somewhat more rapidly, especially in the 1970s. However, the user cost of housing – which better reflects the economic costs of ownership – rose only sluggishly. These relatively modest increases in the price of housing helped to make it more affordable. As much as the rise in nominal incomes, the increasing affordability of housing led to increased consumption over the postwar years.

As noted by Beresford and Rivlin (1966), one of the ironies of prosperity is that it can actually reduce household income. With a higher income, household members who previously could not afford to live alone now separate. If they take an income with them, two households may result, each poorer than its common predecessor. One of the riddles of contemporary housing policy concerns the continued large number of households with affordability problems (i.e. a high ratio of shelter expense to income) at a time when housing costs are rising only modestly relative to per capita income. Part of the reason is this impact of income (or more correctly housing affordability) on household formation itself.

Thus increased household formation and increased housing demand can be explained in part by the same economic forces. Whatever led the price of housing services to rise slowly, and incomes to grow more rapidly, in part caused the formation of more households and more housing consumption. It is beyond the scope of this study to comment in greater depth on this. The demand perspective adopted in this book ignores supply factors that might have shaped the price of housing. Thus I am not in a position to comment, for example, on the role of improved construction productivity in the price of housing.

Finally, this book cannot be concluded without discussing shifting patterns of tenure. Owning and renting are two distinct choices for most consumers. Over the early part of the postwar period, this choice was also between distinctive types of housing, at least in that it was difficult for owner-occupiers to live in apartments. However, the emergence of new forms of ownership (e.g. condominium ownership) largely eliminated this important tenure distinction. What was left in part were the relative economic costs of owning versus renting. In this view, consumers choose to become home owners if the costs of owning are less than renting. Evidence suggests that, during the 1970s, the economic costs of owning fell relative to the costs of renting, principally because of the capital gains attendant upon house price appreciation and the income tax advantages of ownership. This helps to account for the rise of home ownership at that time.

However, tenure choice is not based simply on a calculation of the relative costs of owning versus renting. Housing is more than just another asset. Owning and renting, for example, provide different kinds of security of tenure. Often there is less likelihood of being forced to move if one is a home owner. As a renter, one has always the prospect that the landlord may force a move when one does not want a move. Households that are risk averse, or that face high relocation costs if forced to move, will have a preference for ownership. Households that are greater risk takers, or have low costs of relocation, will be more sensitive to the comparative financial costs of owning and renting. This is an important argument. It helps to explain why families with children, who typically have large psychic and/or monetary costs of moving, are more prone to home ownership. Perhaps, also, risk aversion among the elderly pushes them toward ownership. However, young households without children often have low costs of relocation and are likely to move again soon anyway. As such they are less impelled to ownership.

It has long been recognized that tenure varies among demographic groups. In fact, the data in Tables 81 and 82 suggest that demographic variations were as important as income variations in describing postwar tenure patterns. Commonly, demographic differentials are attributed to either savings or marginal tax rates. Young households typically do not have the savings required to make a substantial downpayment. However, the savings of middle-aged and elderly households are higher; hence they can more easily afford ownership. In effect, this argument asserts that imperfections in the capital market prevent low-savings households from becoming home owners. Higher-income households also face higher marginal tax rates. The user (economic) cost of home ownership declines at higher incomes, principally because of the nontaxation of capital gains and the imputed returns on owner equity. Middle-aged groups tend to have the highest average incomes, thus the lowest typical user cost, hence the greatest likelihood of being owners. Other age groups typically have lower marginal tax rates and hence are less likely to own.

What happened in the postwar period reflects all of these demographic variations. During the baby boom, family sizes increased. In itself, this should have delayed ownership. In part, larger families typically find it more difficult to accumulate savings, given the spending pressures on their incomes. In part, the presence of young children typically disrupted the work-force participation of wives, cutting into their current earnings and possibly lowering their lifetime earnings trajectory. However, the increase in family sizes made security of tenure more important, a desire abetted by the growth in real incomes. During the 1950s and into the 1960s, these two sets of effects largely offset one another. In the 1970s, family size began to fall. Presumably, the need for security of tenure subsided somewhat as a result. However, the growth in real income, in part attributable to increased work-force participation among wives, combined with the impact of capital gains on the user cost of home ownership, tipped the balance in favour of ownership for many consumers. Throughout the postwar period, the tax system increasingly favoured ownership. Prior to the 1970s, tax rates were not indexed, and so typical marginal tax rates increased with inflation and with the growth in real income. In the 1970s, the new Income Tax Act increased the attractiveness of home ownership by making other kinds of capital gains taxable for the first time.

Much of the demographic variation in tenure is attributable to these explanations: security of tenure, marginal tax rate variations, and savings behaviour. However, we do not yet understand very well how savings behaviour varies over the life cycle, how sensitive it is to changes in family size, and how fertility, work-force participation, the income tax structure, and savings behaviour are linked. We have even less information about the importance of security of tenure over the life cycle or the extent to which ownership contributes to it. There may well also be other important reasons why demographic variations exist. However, much of the demographic variation is attributable to these three explanations, and future research in this area should focus on them.

Notes

CHAPTER ONE

1 Cf Dynes (1978) and City of Toronto Planning and Development Department (1980a; 1980b).

CHAPTER TWO

1 It is easy to fall into the trap of thinking of housing as a durable, unmalleable stock. When told that there are 8 million dwellings in Canada, we commonly think of neighbourhoods of detached or attached housing units and apartment buildings. How often do we think of other possible housing forms: such as vans, tents, or houseboats or mobile homes (or even, in the words of the 1941 Census, "a dugout in the ground, if used for human habitation")?

2 The impact of this change in interpretation is substantial when put in perspective. The net increase in occupied (private) dwellings in Canada between 1941 and 1951 was only 833,000 units (and the net change in occupied apartments and flats only 353,000). Thus a change in interpretation could have accounted for as much as 20 per cent of the total increase in dwellings and almost 50 per cent of the increase in the stock of apartments and flats.

3 The census also recognizes "vacant" dwellings. A definition of a vacant dwelling and its relationship to an overall count of the housing stock are presented in chapter seven.

4 Recent American censuses use a similar, but not identical, definition for a (private) dwelling. As explained in Kobrin (1976a, 128): "The important components of [census] household definitions are use (indicated ordinarily by the presence of various items of cooking equipment); privacy (the consideration here is usually separate access); and the number of unrelated persons who are present." The Canadian definition adopted here differs mainly in the "use" attribute. In defining a dwelling (and thus a household), the Canadian defi-

nition looks to whether individuals share rooms or space. The American definition looks to whether individuals have their own cooking facilities.

5 Also, the 1 June date used in censuses is noteworthy here. Most university and college students have finished the academic year by May. Thus students who might be living and "normally resident" elsewhere for eight or nine months of the year are enumerated where they reside for the spring-summer term. 1 June is too early, however, for many cottagers to have opened their summer homes. Retired persons, and others, who spend several months a year at a summer home thus tend to be enumerated at their winter homes. Some of the retired who winter in the south (e.g. Florida) and summer in Canada may not be enumerated at all.

6 In the 1981 Census, each person was asked his or her name, age, sex, marital status, and relationship to "Person 1" (see below for a discussion of this term). A common law relationship is assumed to exist if a person in the household checked off the box marked "Common law partner of Person 1" as his/her relationship to Person 1. The common law relationship was not detected if that person listed the relationship otherwise (e.g. "Room-mate"). Another case where a common law relationship is not identified arises when neither spouse is "Person 1." Prior to 1981, the structure of the questionnaire makes the identification of common law couples virtually impossible. The 1976 Census questionnaire, for example, did not explicitly mention common law relationships at all, although an accompanying guidebook instructed such persons to list themselves as married.

CHAPTER THREE

1 It does not show how many children will typically be borne, because that depends on fertility rates at the time that a woman is in each age group. Consider, as an example, 1,000 women born between 1921 and 1925. After Second World War, these women had more children than would have been expected given the fertility rates of women 20 and over in 1940. In total, they averaged 3,717 births. This latter number is the cohort total fertility rate (CTFR). Of course, we cannot observe a CTFR until after a cohort has substantially finished passing through its child bearing years. This constrains use of the CTFR because some women bear children as late as their forties. Thus CTFRs can often be calculated only for cohorts born at least 40 years ago. Of course, it is possible to guess at the remaining birth expectancies of women who have passed through only part of their child bearing years. Bloom (1982) is an example of such a study.

2 There are separate life tables for men and women. In combining these, I assumed that there were 5 per cent more males than females at birth.

3 These first marriage rates are similar to (but slightly different from) probabilities of single persons marrying in the given year. First marriage probability

schedules for Canada have been estimated by Laing and Krishnan (1976), Mertens (1976), and Adams and Nagnur (1981). These three studies are based on different methods and assumptions, limiting any comparisons of their results.

4 Tsui excluded child births prior to, or within nine months of, marriage.

5 See Abernathy and Arcus (1977) for a more detailed discussion of divorce legislation in Canada.

6 A divorce rate is like, but not identical to, the probability of a married person becoming divorced. The probability of divorce is the ratio of divorces during a year to the population at risk at the start of the year. The divorce rate described in this text used the mid-year population, not the population at the start of the year. Probabilities of divorce are not widely available. For an example of such data, see Adams and Nagnur (1981).

7 As with FMR, these are not the same thing as probabilities of remarriage. Canadian remarriage probabilities for 1961 and 1966 can be found in Kuzel and Krishnan (1973) and for 1975-7 in Adams and Nagnur (1981).

CHAPTER FOUR

1 The formula for doubling time (t) in year is: $t = ln(2) / ln(1 + r)$ where r is the compounded annual rate of growth in decimal form, and ln is the natural logarithm.

2 Data taken from Kobrin (1976a, 129) and Population Reference Bureau (1982, 13).

3 Sweet (1984, 131) notes an almost identical increase in nonfamily households in the United States, from 19 per cent of all households in 1970 to 26 per cent in 1980.

4 Note that these men cannot head a husband-wife family because they are currently widowed. This emphasizes the fact that the census looks to the current marital status of the individual; widowers who remarry are defined to be married. It also reflects the census treatment of common law partners as married, regardless of the stated marital status.

5 I am not urging reinstatement of the clearly outmoded pre-1976 definition. Rather, I am concerned about the difficulty of historical comparability when definitions change. If anything, the problem is exacerbated by the switch in the 1981 Census to the "household maintainer" concept.

6 Some American evidence by Michael et al (1980, 46-9) supports this conjecture, albeit weakly. Michael et al examined interstate differences in the propensity for widows (65 or older) to live alone in 1970. The dependent variable was the proportion of widows living alone. One of the independent variables was the ratio of women 65 or older to the number of women 35-44. They refer to this as the "mother-daughter" ratio, although in- and out-migration can play havoc with such an assumption. They find that widows are more likely to live alone in states where this "mother-daughter" ratio is larger. This substan-

tiates Kobrin's argument. However, Michael et al find that the empirical relationship is weak and not statistically significant.

CHAPTER FIVE

1 A brief description of the public use samples is enlightening. These are 1 percent random samples of individual census returns. In other words, for one person in every hundred, we are given that individual's responses to most questions on the census. In creating these samples, Statistics Canada excluded, edited, or recoded, certain information to preserve the anonymity of the respondents and to "fill in" missing data. Also to preserve anonymity, observations in lightly populated Prince Edward Island, Yukon, and the Northwest Territories were excluded. However, for the most part, these samples are smaller replicas of the original census data. Outside direct access to the original census returns, these are the largest and most detailed samples of households in Canada currently available.
2 The three samples (individual, family, and household) are similar in structure, though not exactly comparable. It is possible, for example, to identify in the individual sample those persons who are household heads. The total of these is very close to, but not the same as, the number in the household sample. Similarly, the number of individuals reported as family heads (in the individual sample) is slightly different from the number in the family sample. There are two reasons for this. One is that drawing of the three random samples is done independently, so there is no guarantee that comparable counts will in fact be identical. Second, there are small differences in the way that the universe for each sample is defined. However, it should be emphasized again that the differences are small in practice. Further, with access to the full census data base, the correspondences could be made exact.
3 Distinguishing between domestic individuals and nonresidents is straightforward in the 1971 public use sample (individual file). However, the 1976 sample lumps nonresidents together with some other categories, e.g. temporary residents who could not be assigned to the dwelling in which they normally resided in Canada and persons for whom relationship to head of household at the usual place of residence could not be determined. This does introduce a comparability problem. However, the numbers of individuals thus excluded from the domestic population are not large.
4 The 1976 Census public use sample (family file) does not identify family headship in the case of husband-wife families. In calculations for this chapter, the husband is assumed to be the family head in 1976 unless the wife is identified as the household head. This rule-of-thumb overstates the propensity for a husband to be family head to the extent that wives can also head lodging (i.e. nonprimary) husband-wife families. In 1971 Census data, the husband (if present) is always taken to be the family head.

5 Table 37 omits propensities that have no direct effect on the formation of one-person households by definition. This includes $p[5]$, $p[6]$, $p[13]$, $p[15]$ through $p[29]$, and $p[32]$ through $p[35]$.

CHAPTER SIX

1 Steele used a logit model like the one described later in this chapter. In her analysis, the dependent variable is the propensity of a married man and his immediate family to live alone. This is similar to the product of $q[1]$ and $q[2]$ above. Age of husband, period of immigration, mother tongue, labour force status, level of education, and income were the explanatory variables.

2. Glick (1976) reports that, in the United States in 1970, this option was common only among women under 25. Sweet (1972), in studying estranged women in the 1960 U.S. Census, finds this option prevailed only among women with just one child, and only where the child was very young (typically under three years old).

3 He does not seem to be impressed by his own findings, though, and blames his results on problems with the census definition and measurement of income. In the 1960 U.S. Census questionnaire, no explicit mention is made of alimony or child support, for example. Respondents may very easily have omitted such amounts in estimating their "other" income. Sweet also notes that some respondents may have under-reported their income for a variety of reasons. Sweet also mentions the problem of using last year's income to look at current living arrangements.

4 This proportion was their dependent variable. They employed linear and logit models to estimate the impact on this proportion of certain independent variables. For 25–34-year-old singles, the independent variables included average statewide (i) per capita income, (ii) index of liberalness, (iii) rate of recent in-migration, and (iv) level of education. For elderly widows, the independent variables were similar, except that a "mother-daughter" ratio was substituted for (ii). Michael et al found that these variables accounted for much of the interstate variation in propensity to live alone. Further, although estimated using only 1970 data, the models came close in predicting the aggregate rate of American one-person household formation between 1950 and 1976.

5 Following the approach of chapter five, I ignore census respondents enumerated overseas or in collective dwellings. The numbers are small in any case.

6 Such support could take a number of different forms. It could be a periodic cash transfer (e.g. assisting monthly rent payments) or a lump sum payment (e.g. a downpayment on a first home). Alternatively, the assistance might be in the form of a reduced rate, interest-free, or forgivable loan. In other cases, the assistance might be an "in kind," rather than a cash, transfer. Baby-sitting services, or the use of an automobile or cottage, would be such instances.

7 Families that maintain a dwelling are a subset of all families. Hence the sample sizes are somewhat smaller than in Table 40. Also, primary families typically have a higher income than do lodging families. Hence the mean incomes reported in the second column of Table 41 are larger than those in Table 40.

8 Only among elderly lone parents were they comparable. Remember that the Census includes, as lone parents, anyone coresiding with never-married children, regardless of age. A 74-year-old woman could well be living with a "child" aged 50. In such cases, the principal contributor to family income likely was the child. Thus it is not surprising that, among the elderly, two-person lone-parent families had higher family incomes than husband-wife units without children.

9 To be fair, the latter includes the incomes of both partners, whereas the former is for just one person. However, even taking this into account, nonfamily men had lower incomes on average than did their married counterparts.

CHAPTER SEVEN

1 Another illustration of the importance of conversions is found in the *1951 Census of Canada*, vol 10, 361. There it is reported that between 1941 and 1951 the number of single attached dwellings (i.e. semi-detached and row houses) in Hull, Quebec, fell from 1,500 to 345. A subsequent survey indicated that roughly one-third of the dwellings reported as single attached in 1941 fell into the same category in 1951. However, the remaining two-thirds had been converted into other categories, mainly apartments and flats.

2 The censuses of 1951, 1961, and 1966 also include dwellings under construction in the total housing stock. Since 1971, this category has been excluded. For consistency, the housing stock estimates presented in this chapter all exclude units under construction.

3 During the prosperous 1920s, Firestone estimated that the annual increase averaged 2.2 per cent, still below the postwar average. The estimated increase dropped during the 1930s to 1.2 per cent, and climbed slightly during the war years, to 1.4 per cent. It rose again, in the early postwar years, to 2.5 per cent between 1946 and 1949.

4 In making this observation, note that the definition of rural and urban areas changed over the years. In the 1941 Census, all areas within the boundaries of incorporated cities, towns, and villages were urban, and the rest of the country was rural. In the 1951 Census, urban areas included only cities, towns, or villages of 1,000 population or more, whether incorporated or not. Also included as urban were any areas lying within the boundaries of a census metropolitan area, some of which would previously have been treated as rural. In the 1956 and 1961 censuses, the definition of urban was expanded to include the urbanized fringe of smaller urban centres. Urban areas were defined in the same way in 1966 and 1971 as in 1961. In 1976, the definition

was revised again: "The definition of urban area is more restrictive in 1976 . . . All areas must now show a density of 1,000 or more persons per square mile (386 per square kilometre), as well as a minimum population of 1,000, to be classified as urban. In previous censuses, the density applied only to areas outside the incorporated cities, towns, and villages. While there was a significant reduction in area classified as urban because of the new definition, the reduction in population so classified was minimal" (*1976 Census of Canada*, vol 1.1, 28-9). The 1981 Census used a similar definition except that the density requirement was raised to 400 persons per square kilometre.

5 Of course, it must be remembered here that "increasing urbanization" is just a proxy for the variety of demographic, social, and economic differences between rural and urban areas. To the extent, for example, that urbanites are more or less likely to marry, have children, or divorce, to the extent that their incomes differ from those of ruralites, or to the extent that their preferences are different, they may well choose different living arrangements and housing forms.

6 In the 1951 Census, the dwelling had to be owned by the head or a member of the immediate family. In principle, the 1951 definition could be more restrictive because it more narrowly limits who the owner can be. However, there likely was little difference in practice because it was uncommon for a dwelling to be owned by a person who was a household member but not its head or an immediate relative.

7 I am not aware of a direct estimate of the extent of foreclosure-based ownership in the 1930s or early 1940s. Indirect evidence in Firestone (1951, 274), though, suggests that this may have been substantial. Firestone estimates that between 1931 and 1933 the number of owner-occupied dwellings in Canada declined in net by 18,400, or about 1.3 per cent. Many of these represent foreclosures that resulted in the dwelling being left vacant or rented out. Remember, too, that these are net figures; they are the difference between gross additions to owner-occupancy and gross subtractions (which include foreclosures leading to vacancy or tenure change). The gross effect of foreclosures would have been substantially larger.

8 Some of the latter may simply be arising because of errors in dwelling type enumeration in the 1971 Census, e.g. apartments, with their lower incidence of ownership, being misclassified as single, attached dwellings.

9 There is little information available on the relative popularity of wood frame and masonry wall construction. CMHC (1958, vi-4) reports that, at that time, about three wood frame dwellings were being constructed for every masonry wall dwelling.

10 I am grateful for the clarifying comments of Morris Clayton of Canada Mortgage and Housing Corporation on this issue.

11 Although there are no estimates of average age of dwelling before 1921, there are estimates of the volume of new residential construction. Saywell (1975, 4-

13) reviews the work of Steele and Pickett and concludes that, during the boom of 1905-10, housing starts reached 8.7 per thousand population. This was a rate not to be exceeded, or even approached, until the late 1960s.

CHAPTER EIGHT

1 The components making up the bundle, and their weights, are described in Table 67.

2 Shelter expense accounted for 20.7 per cent of consumer expenditure in 1978, and household operation just 8.7 per cent. Shelter expense included rented accommodation, owned accommodation, and a residual "other" category (a minuscule 0.5 per cent of consumer expenditure).

3 There is not much published information on the calculation of the price index for shelter. The description here is based on the sources listed in Table 67.

4 This point was recognized by Statistics Canada at an early date: "Changes in rents of new rental units are recorded from the time such units are introduced into the sample, but the difference in the absolute level of rents between the old and new units is not taken into account in constructing the rent index. This procedure of introducing rents for new units so that they do not affect the movements of the index, has been used because higher rents for new units are not comparable to rents of older dwellings" (Statistics Canada. 1952. *The Consumer Price Index:* January 1949-August 1952. Catalogue 62-502, p. 15).

However, the immediate postwar period marked the end of wartime rent controls, and rents jumped sharply. SCRI would have missed such price increases because of the way it is estimated. To correct for this, the following procedure was used (p. 15): "During the latter period of federal rent control, units were freed from control as they became vacant. Consequently, within the same apartment block, dwelling units having the same number of rooms and facilities, were commanding different rents . . . A survey of such apartments showed that in June 1952, decontrolled apartments were renting for an average of twenty per cent more than identical but controlled apartments. This figure was accepted as a measure of . . . the difference between rents of new and other rental units attributable to rent control . . . and has been applied to new urban rental accommodations built since the base year 1949 . . . By August 1952 it had added 1.1 points . . . to the unadjusted rent index."

5 Statistics Canada has two additional price measures: the input price index for residential construction (IPIRC), and the implicit price deflator for residential gross fixed capital formation (IPDRGFCF). IPIRC measures material prices and rates of labour employed in single detached residential construction. IPDRGFCF is derived from the ratio of current to constant dollar estimates of total gross additions to capital, excluding land costs, for all types of housing. Unlike

NHPI, neither index includes land costs. Also, IPIRC is restricted to just one kind of new housing (i.e. single detached units), while IPDRGFCF includes dwelling extensions and other forms of residential fixed capital formation besides just new housing.

6 Other elements in the cost of housing, such as expenditures on fuel, electricity, and water, are included in the "Household Operation" component of the CPI. These are included in SCHI, but not in SCOI.

7 However, with a long-term lease and with the right to sublet, the tenant could possibly realize a financial gain from subletting. This is effectively possible only when the least amount comes to be substantially below the current market rent. In such cases, the rent a lessee might be willing to pay would take account of the future gain associated with a sublet.

8 The mortgage interest component is calculated as a moving average of interest rates applied to the principal remaining on a fixed stock of dwellings. The amortization and percentage equity contribution of owners are assumed to remain constant. As described in Chambers (1983, 240): "a) The principal outstanding in the applicable base family expenditure survey period distributed over 60 months is revalued on a moving average basis each month by . . . the New Housing Price Index (inclusive of land) . . . This produces a fractional distribution over 60 months of principal outstanding valued at the new house purchase price . . . b) Average mortgage rates are calculated for each of the 60 months in the moving average. c) For each of the 60 months, the average mortgage rate is applied to the fraction of the principal outstanding."

9 Depreciation is assumed to be a fixed percentage of dwelling value. The CPI weight assigned to replacement is 2 per cent of the value of the owner-occupied housing stock as estimated from the base year family expenditure survey. In subseqeunt years, Statistics Canada's residential building construction input price index was used to update the replacement cost index (before the 1974 revisions to the CPI). With the 1974 revisions, the new housing price index was used.

10 The absence of national data on house price changes forced McFadyen and Hobart (1978) to estimate G using a price index for new houses financed under the National Housing Act (NHA). They comment (p. 109) on the discrepancies between a true quality-adjusted house price index and the NHA index: "First, as urban centres expand, residential construction will take place on [more distant] building lots . . . This decrease in location quality will be reflected in NHA dwelling prices. Second, in recent years NHA-financed dwellings have become the lowest-price segment of new housing construction because of limits placed on the maximum NHA loan. Also [many] NHA dwellings have in recent years been constructed [in] areas . . . characterised by lower rates of price increase . . . If dwelling quality has increased over time with quality being reflected in rising prices, the result will be an upward bias in the mea-

surement of capital gains: if dwelling quality has decreased the converse would hold."

11 In earlier years, Statistics Canada used the residential building construction input price index (RBCIPI) instead of NHPI. RBCIPI was based on a fixed bundle of construction input materials and labour that, like NHPI, can be viewed as a quality-controlled price index. Over the entire period, the price index for owners' repairs expense was also based on RBCIPI.

12 Throughout this period, women earned substantially less than men. Thus, even though female participation rates increased, the incomes of working wives typically remained far below those of their husbands. The following data are the ratios of the average income of wives (with income) to the average income of male household heads (with income) as reported in Statistics Canada's *Income Distribution by Size in Canada* for various years:

1951	1957	1961	1965	1971	1976	1981
0.34	0.36	0.37	0.34	0.37	0.38	0.45

Only in 1981 was there any evidence of a substantial improvement in wives' earnings.

13 An economic family includes all persons coresiding in a household who are related by blood, adoption, or marriage. An unattached individual is a person living alone, or living with one or more other persons with whom he or she does not form an economic family. In most cases, a household is made up of just one spending unit. In the others, a household consists of two or more.

14 In interpreting this series, note that, prior to 1966, these average incomes were for nonfarm spending units only. Subsequently, the averages cover both farm and nonfarm units.

15 It is difficult to make comparisons between AUY and PCPDI, however, because the difference between them reflects both changes in living arrangements and changes in the difference between gross income and disposable income. The latter difference is sensitive to changes in taxation. Changes in income tax rates and the introduction of public pension and medical insurance premiums substantially altered the relationship between gross and disposable income over the postwar period.

16 Bartlett and Poulton-Callahan (1982) report a similar finding. They compare U.S. income distributions in 1951 and 1976 for all types of families. Of the four family types - female-headed, male-headed without wife, male-headed with nonworking wife, and male-headed with working wife - the last traditionally had the most equitable distribution of income. As wives increasingly entered the labour force, the authors foresee a decline in the inequality of distribution of family incomes. In a study of British couples, Layard and Zabalza (1979, s137) come to a similar conclusion about the effect of working wives on income inequality: "Interestingly, there is almost no correlation between hus-

band's and wife's earnings . . . Although husband's and wife's wages are positively correlated, the effect of this is offset by the negative correlation between husband's wages and wife's annual hours."

17 There is a small comparability problem here in that the 1951 and 1961 data on unattached individuals are for the nonfarm population alone, whereas the household data are for all Canada. Thus, the 1951 and 1961 figures understate the number of unattached individuals in all Canada. No adjustment has been made for this. However, the trend described in the text would be even more evident if the counts were adjusted.

18 Hickman does not make clear just how HCCA is calculated. Presumably, the numerator is average expenditure on new residential construction. The denominator must be a price index of some kind. Hickman refers to HCCA simply as a "constant 1958 dollar" expenditure.

19 The above two paragraphs rest on some assumptions about 1946: that there were 2.9 million households in Canada; that the private population was about 11.5 million persons; and that average household size was therefore about 4.2 persons.

20 There are two versions of Smith's model: an aggregate version and a version where single and multiple dwellings are treated separately. For purposes of the present discussion they are quite similar. To simplify the presentation, I consider only the aggregate version.

21 The handling of the consuming unit is of interest here. Hickman uses either the actual number of households (HH) or the number of households expected using 1940 headship rates (HHS). Smith uses the number of census families. Muth uses total population in the denominator of his demand variable. In each case, what we want is a measure of the potential number of consuming units. Presumably this is more than the number of families (since some nonfamily individuals will form separate households), but less than the number of people (since it is unlikely at any income or price that every person would live alone). The current number of households indicates the number of consuming units currently but does not take into account its elastic nature given income or price changes.

22 I don't think that the absence of small, owner-occupied dwellings was a substantial problem for owners. In the early postwar period, for example, there was much construction of small owner-occupied dwellings in response to a perceived market for low-cost housing. However, there were some restrictions on the ability of developers to offer small units. Some municipalities had building-code or zoning provisions that prevented construction of small units. Also, before the late 1960s - when the condominium form of ownership was legalized in much of Canada - it was not easy to create shared ownership of apartment buildings. It is unclear just how important such restrictions were in forcing consumers who desired small dwellings, or did not want a yard, to rent rather than own.

23 For renters, it is simply the price discount described above. For owners, price is defined to be the product of the price discount and the ratio of housing expense (including mortgage interest, taxes, fuel, utilities, and repairs) to dwelling market value.

24 Other possible explanations seem weak. SCHI may have understated the postwar price change (principally because of the exclusion of new rental dwellings). Faster-rising house prices would have helped offset the decline in the housing budget share. However, SCHI could either understate or overstate the actual change in housing price. It is not clear that SCHI is necessarily biased downward. Another explantion is that the constant budget share resulted from a changing demographic composition. More specifically, the rise of the person living alone with his or her typically higher budget share may have caused housing expenditure to keep pace with income overall, in spite of the low income elasticity. However, this demographic shift likely was not large enough in magnitude to offset the substanital negative impact on budget share of rising real income. For example, if the income elasticity was 0.3, and real income doubled over the postwar period, (35) suggests that the housing budget share would have declines by over one-third. It is unlikely that the rise of the person living alone (from 7 per cent of all households in 1951 to 20 per cent in 1981) was enough to offset this. Another conjecture is that cross-sectional elasticity estimates, being based only on 1970s data, are unrepresentative of the entire postwar period. In other words, the elasticities may have been larger in the 1950s and 1960s. This could explain why the housing budget share did not decline. However, I suspect that income elasticities were not much higher in previous decades. Virtually every cross-sectional study available, both for the 1970s and earlier, suggests that high-income households spend a relatively smaller portion of their budget on housing. This is consistent with low-income elasticity. There is no compelling reason why income elasticities would have been substantially higher in these earlier times.

CHAPTER NINE

1 This is the same War Measures Act used to suspend civil liberties in Canada in October 1970.

2 Cf Saywell (1975, 213).

3 This description of the loan scheme is adapted from Jones (1978).

4 In an effort to increase the supply of NHA-insured mortgage money, the government over the years kept altering the maximum permissible interest rate. For a time, in the late 1960s, it was tied to the government's long-term bond rate. In 1969, the interest rate limit was removed entirely.

5 Saywell (1975, 205).

6 Cf Central Mortgage and Housing Corporation (1971, 21).

7 The disadvantage to the lender is the risk that a roll-over mortgage may not be available.

8 Although Murray's findings are interesting, the crudeness of the model should not be forgotten. A subsidy that lowers the price of home ownership has two effects. First, it encourages some consumers who would have purchased anyway to purchase now under the program. The only rational consumers who would not seek subsidized units would be those who were restricted from participating: typically, either they are ineligible (e.g. if their income was too high) or they do not want the housing being produced under the program (e.g. if only small dwellings were eligible). A better formulation of the model would consider such regulatory constraints. Second, the subsidy encourages some households to become owners that otherwise might have remained renters or not become owners until a later date. A large subsidy per unit household might induce many such new owners. A small subsidy would induce fewer. Therefore it is important to specify in the model the size of the typical subsidy provided. Murray's model does not do this. It is therefore difficult to know just how responsive demand might be in other jurisdictions, such as Canada, where the terms of the ownership subsidy are different.

9 More specifically, the government did not guarantee each and every loan. It insured each approved lender to the extent of 15 per cent of such loans, up to an aggregate amount of $50 million. The program was terminated in 1940, when this limit was reached. See Firestone (1951, 481).

10 Another program was the Emergency Shelter scheme, involving the use of surplus wartime huts as housing of last resort. It was used primarily to house married students attending university in the early postwar years.

11 Loans were also made, under certain circumstances, to landlords, to nonprofit housing corporations, and to co-operatives to improve the rental housing stock.

12 Taken from the *Annual Report* of CMHC, 1978 (p. 28).

13 The estimates presented in this paragraph are taken from Dowler (1983, 22).

14 The increase in home ownership in the United States was much more marked than in Canada. In speculating on why, one possible argument is a difference in demographic structure between the two countries. However, Rosen and Rosen do not include any demographic variables in their tenure choice model. They did attempt to include demographic variables but found them to be statistically insignificant. Given the upward and downward movements in the incidence of home ownership in Canada, and the emphasis put in chapters 7 and 8 on demographic explanations, Rosen and Rosen's finding is all the more puzzling.

15 Firestone (1951, 484).

16 Central Mortgage and Housing Corporation (1971, 19) and Canadian Council on Social Development (1977, 3).

17 Datum taken from *Canadian Housing Statistics, 1982*, 54.

18 Datum taken from *Canadian Housing Statistics, 1980*, 53.

19 In the Revised Statutes of Canada, 1971, section 35A became section 40, and 35D became 43.

20 Refer to *Canadian Housing Statistics*, 1955.

21 Refer to *Canadian Housing Statistics*, 1955 and 1964.

22 Central Mortgage and Housing Corporation (1971).

23 Dowler (1983, 24).

24 This is a point made, for example, in Dowler (1983, 25) and L.B. Smith (1983, 61).

25 Dowler (1983, 43).

26 In addition, the landlord had the possibility of postponing capital gains taxation through "roll-over" provisions, provided that the gains were reinvested in other rental buildings.

27 Goldberg (1980) examines the relative costs of "goldplated" versus "minimal" servicing standards as used in the cities of Vancouver and Richmond, British Columbia. His evidence shows that, over the long run, it was more efficient to provide high initial servicing standards than to upgrade initially minimal standards.

28 Comay Planning Consultants et al (1973, 147).

29 Under the province's Planning Act, large parcels of land could be severed, without subdivision approval, under a "consent" procedure. Consents were granted by local governments, and the practice varied from one locality to the next. In the northern fringe of Metropolitan Toronto, for example, it was not uncommon for farmland to be severed into lots as small as 4 hectares (10 acres) using the "consent" procedure. Below that size, the province generally insisted on subdivision approval.

30 Proudfoot (1980, 80) does discuss the time delay associated with rezoning applications in the City of Toronto. However, he does not estimate the resulting cost to the developer nor the subsequent impact on the price of housing constructed on those properties.

31 Ontario has revised its original Planning Act of 1946 on several occasions. The 1983 act is described here simply to illustrate the range of concerns in provincial legislation across Canada. The reader should remember that the actual provisions of the Ontario legislation did vary somewhat over the postwar period.

32 An example of this was the attempt by the City of Toronto, as well as other municipal governments, to rid itself of body rub and massage parlours and other "adult entertainment facilities." These "land uses" had proliferated in downtown areas during the 1970s. Several cities, including Toronto, passed zoning bylaw amendments that restricted such uses to areas of the city zoned "Heavy Industrial." The intent was to at least get these facilities out of the downtown area and ideally to eliminate them altogether by forcing them into locations that were inaccessible and therefore unprofitable. While this exam-

ple does not have a direct effect on residential construction, it is an example of zoning that does not provide a "reasonable selection of comparable alternative locations."

33 This is discussed in Hutcheon (1971, 4).

34 I am not trying to argue here that building codes are needlessly expensive. Proponents of building codes argue that the incremental cost is small and that the buildings constructed are, in fact, of better quality.

References

Abernathy, T.J., and M.E. Arcus. 1977. The law and divorce in Canada. *Family Coordinator* 26: 409–13.

Adams, O.B., and D.N. Nagnur. 1981. *Marriage, Divorce, and Mortality: A Life Table Analysis for Canada, 1975–1977*. Catalogue 84-536. Ottawa: Statistics Canada.

Bartlett, R.L., and C. Poulton-Callahan. 1982. Changing family structures and the distribution of family income: 1951 to 1976. *Social Science Quarterly* 63: 28–38.

Basavarajappa, K.G. 1978. *Marital Status and Nuptiality in Canada*. 1971 Census of Canada Profile Studies, Vol 5, Part 1. Catalogue 99-704. Statistics Canada.

Beaujot, R. 1977. Components of change in the numbers of households in Canada, 1951–1971. *Canadian Journal of Sociology* 2: 305–19.

Becker, G. 1960. An economic analysis of fertility. In National Bureau of Economic Research, *Demographic and Economic Changes in Developed Countries*, 209–40. Princeton: Princeton University Press.

Beresford, J.C., and A.M. Rivlin. 1966. Privacy, poverty, and old age. *Demography* 3: 247–58.

Bernard, J. 1975. Note on changing life styles, 1970–1974. *Journal of Marriage and the Family* 37: 582–93.

Blake, J. 1974. Can we believe recent data on birth expectations in the United States? *Demography* 11: 25–44.

Bloom, D.E. 1982. What's happening to the age at first birth in the United States? A study of recent cohorts. *Demography* 19: 351–70.

Brady, D.S. 1958. Income and consumption: individual incomes and the structure of consumer units. *American Economic Review*, 48 (Supplement): 269–78.

Canadian Council on Social Development. 1977. *A Review of Canadian Social Housing Policy*. A study by the Canadian Council on Social Development. Ottawa: Canadian Council on Social Development.

Carlson, E. 1979. Divorce rate fluctuation as a cohort phenomenon. *Population Studies* 33: 523-36.

Carter, H., and P.C. Glick. 1976. *Marriage and Divorce: A Social and Economic Study*. 2nd edition. Cambridge: Harvard University Press.

Central Mortgage and Housing Corporation. 1955. *Building Standards (Excluding Apartment Buildings)*. Ottawa: CMHC.

–1958. *A Review of Housing in Canada. A Brief from Central Mortgage and Housing Corporation to the United Nations*. Ottawa: CMHC.

–1971. *Housing in Canada: 1946–1970*. A supplement to the 25th annual report of CMHC. Ottawa: CMHC

Chambers, E.J. 1983. Recent comparative trends in the Canadian and U.S. CPIs: the treatment of homeownership. *Canadian Public Policy* 9: 236-44.

Cherlin, A. 1977. The effect of children on marital dissolution. *Demography* 14: 265-72.

Chevan, A., and J.H. Korson. 1972. The widowed who live alone: an examination of social and demographic factors. *Social Forces* 51: 45-53.

City of Toronto Planning and Development Department. 1980a. Housing deconversion: why the city of Toronto is losing homes almost as fast as it is building them. *Research Bulletin* 16. Toronto: City of Toronto Planning and Development Department.

–1980b. Households in new units: an analysis for the city of Toronto, 1974-77. *Research Bulletin* 17.

Comay Planning Consultants et al. 1973. *Subject to Approval: A Review of Municipal Planning in Ontario*. Toronto: Ontario Economic Council.

David, M.H. 1962. *Family Composition and Consumption*. Amsterdam: North Holland Publishing

Dennis, M., and S. Fish. 1972. *Programs in Search of a Policy: Low Income Housing in Canada*. Toronto: A.M. Hakkert.

Diamond, D.B. 1980. Taxes, inflation, speculation, and the cost of homeownership. *Journal of the American Real Estate and Urban Economics Association* 8: 281-98.

Divic, A. 1982. Housing Requirements, Housing Demand and Housing Supply in Canada: 1976-1981 . Manuscript. Market Forecasts and Analysis Division, Canada Mortgage and Housing Corporation.

Doling, J. 1976. The family life cycle and housing choice. *Urban Studies* 13: 55-8.

Dowler, R.G. 1983. *Housing-Related Tax Expenditures: An Overview and Evaluation*. Major Report No. 22. Toronto: Centre for Urban and Community Studies, University of Toronto.

Dynes, S. 1978. Population: a question of balance. *City Planning* 1: 6-7.

Easterlin, R.A. 1980. *Birth and Fortune*. New York: Basic Books.

Fallis, G. 1980. *Housing Programs and Income Distribution in Ontario*. Published for the Ontario Economic Council by the University of Toronto Press. Toronto: University of Toronto Press.

Firestone, O.J. 1951. *Residential Real Estate in Canada*. Toronto: University of Toronto Press.

Glick, P.C. 1976. Living arrangements of children and young adults. *Journal of Comparative Family Studies* 7: 321-32.

Glick, P.C., and A.J. Norton. 1977. Marrying, divorcing, and living together in the U.S. today. *Population Bulletin*, 32.

Glick, P.C., and G.B. Spanier. 1980. Married and unmarried cohabitation in the United States. *Journal of Marriage and the Family* 42: 19-30.

Goldberg, M.A. 1980. Municipal arrogance or economic rationality: the case of high servicing standards. *Canadian Public Policy* 6: 78-88.

Gottlieb, Manuel. 1976. *Long Swings in Urban Development*. National Bureau of Economic Research. New York: Columbia University Press.

Greven, P.J. 1972. The average size of families and households in the province of Massachusetts in 1764 and in the United States in 1790: an overview. In P. Laslett and R. Wall (eds), *Household and Family in Past Time*, 545-60. Cambridge: Cambridge University Press.

Harrison, B.R. 1981. *Living Alone in Canada: Demographic and Economic Perspectives*. Catalogue No. 98-811. Ottawa: Statistics Canada.

Hendershott, P.H., and J.D. Shilling. 1982. The economics of tenure choice, 1955-1979. *Research in Real Estate* 1: 105-33.

Heuser, R.L. 1976. *Fertility Tables for Birth Cohorts by Color: United States, 1917-1973*. United States Department of Health, Education, and Welfare. Washington: CDHEW HRA 76-1152.

Hickman, B.G. 1974. What became of the building cycle? In P.A. David and M.W. Reder (eds), *Nations and Households in Economic Growth*. New York: Academic Press.

Hutcheon, N.B. 1971. *Codes, Standards, and Building Research*. Ottawa: Division of Building Research, National Research Council.

Jones, A.E. 1978. *The Beginnings of Canadian Government Housing Policy, 1918-1924*. Occasional Paper No. 1. Ottawa: Centre for Social Welfare Studies, Carleton University.

Kobrin, F.E. 1973. Household headship and its changes in the United States, 1940-1960, 1970. *Journal of the American Statistical Association* 68: 793-800.

– 1976a. The fall in household size and the rise of the primary individual in the United States. *Demography* 13: 127-38.

– 1976b. The primary individual and the family: changes in living arrangements in the United States since 1940. *Journal of Marriage and the Family* 38: 233-9.

Kuzel, P., and P. Krishnan. 1973. Changing patterns of remarriage in Canada, 1961-1966. *Journal of Comparative Family Studies* 4: 215-24.

Laing, L., and P. Krishnan. 1976. First-marriage decrement tables for males and females in Canada, 1961-1966. *Canadian Review of Sociology and Anthropology* 13: 217-28.

Laslett, P. 1969. Size and structure of the household in England over three centuries. *Population Studies* 23: 199-23.

Laslett P., and R. Wall (eds), 1972. *Household and Family in Past Time*. Cambridge: Cambridge University Press.

Layard, R., and A. Zabalza. 1979. Family income distribution: explanation and policy evaluation. *Journal of Political Economy* 87: s133-s161.

Levine, D. 1977. *Family Formation in an Age of Nascent Capitalism*. New York: Academic Press.

Lindert, P.H. 1978. *Fertility and Scarcity in America*. Princeton: Princeton University Press.

Lithwick, I. 1978a. An Evaluation of the Federal Assisted Home Ownership Program (1976). Ottawa: Unpublished CMHC report.

— 1978b. An Evaluation of the Federal Assisted Rental Program (1976-77). Ottawa: Unpublished CMHC report.

Markandya, A. 1983. Headship rates and the household formation process in Great Britain. *Applied Economics* 15: 821-30.

McFadyen, S., and R. Hobart. 1978. An alternative measurement of housing costs and the Consumer Price Index. *Canadian Journal of Economics* 11: 105-13.

McKie, D.C., et al. 1983. *Divorce: Law and the Family in Canada*. Ottawa: Statistics Canada.

McVey, W.W., and B.W. Robinson. 1981. Separation in Canada: new insights concerning marital dissolution. *Canadian Journal of Sociology* 6: 353-66.

Mertons, W. 1976. Canadian nuptiality patterns: 1911-1961. *Canadian Studies in Population* 3: 57-71.

Michael, R.T., et al. 1980. Changes in the propensity to live alone: 1950-1976. *Demography* 17: 39-53.

Miron, J.R. 1982. *The Two-person Household: Formation and Housing Demand*. Research Paper No. 131. Toronto: Centre for Urban and Community Studies, University of Toronto.

— 1984. *Housing Affordability and Willingness to Pay*. Toronto: Research Paper No. 154. Toronto: Centre for Urban and Community Studies, University of Toronto.

Modell, J. and T.K. Hareven. 1973. Urbanization and the malleable household: an examination of boarding and lodging in American families. *Journal of Marriage and the Family* 35: 467-79.

Modell, J., et al. 1978. The timing of marriage in the transition into adulthood: continuity and change, 1860-1975. *American Journal of Sociology* Supplement 84: s120-s150.

Morrison, P.S. 1978. Residential Property Conversion: Subdivision, Merger, and Quality Change in the Inner City Housing Stock, Metropolitan Toronto, 1958 to 1973. PhD dissertation. Toronto: Department of Geography, University of Toronto.

Murray, M.P. 1983. Subsidized and unsubsidized housing starts: 1961-1977. *Review of Economics and Statistics* 63: 590-7.

Muth, R.F. 1960. The demand for non-farm housing. In A.C. Harberger (ed), *The Demand for Durable Goods*, 29-96. Chicago: University of Chicago Press.

Myers, D. 1978. A new perspective on planning for more balanced metropolitan growth. *Growth and Change* 9: 8-13.

Peiser, R.B. 1981. Land development regulation: a case study of Dallas and Houston, Texas. *American Real Estate and Urban Economics Association Journal* 9: 397-417.

Population Reference Bureau. 1982. U.S. population: where we are: where we're going. *Population Bulletin* 37(2).

Preston, S.H., and J. McDonald. 1979. The incidence of divorce within cohorts of American marriages contracted since the Civil War. *Demography* 16: 1-25.

Proudfoot, S.B. 1980. *Private Wants and Public Needs: The Regulation of Land Use in the Metropolitan Toronto Area*. Working Paper No. 12. Regulation Reference. Ottawa: Economic Council of Canada.

– 1981. The politics of approval: regulating land use on the urban fringe. *Canadian Public Policy* 7: 284-96.

Ritchie, T., et al. 1967. *Canada Builds: 1867–1967*. Toronto: University of Toronto Press.

Rodgers, R.H., and G. Witney. 1981. The family life cycle in twentieth century Canada. *Journal of Marriage and the Family* 43: 727-40.

Rose, A. 1980. *Canadian Housing Policies: 1935–1980*. Toronto: Butterworth.

Rosen, H.S., and K.T. Rosen. 1980. Federal taxes and homeownership: evidence from time series. *Journal of Political Economy* 88: 59-75.

Ryder, N., and C. Westoff. 1972. Wanted and unwanted fertility in the United States: 1965 and 1970. In C. Westoff and R. Parke (eds), Commission on Population Growth and the American Future, Vol 1: *Demographic and Social Aspects of Population Growth*, 471-87. Washington: U.S. Government Printing Office.

Saywell, J.T. 1975. *Housing Canadians: Essays on the History of Residential Construction in Canada*. Discussion Paper No. 24. Ottawa: Economic Council of Canada.

Shorter, E. 1977. *The Making of the Modern Family*. New York: Basic Books.

Smith, J.P. 1979. The distribution of family earnings. *Journal of Political Economy* 87: s163-s192.

Smith, J.P., and M.P. Ward. 1980. Asset accumulation and family income. *Demography* 17: 243-60.

Smith, L.B. 1971. *Housing in Canada: Market Structure and Policy Performance*. Research Monograph No. 2; Urban Canada: Problems and Prospects. Ottawa: Central Mortgage and Housing Corporation.

– 1974. *The Postwar Canadian Housing and Residential Mortgage Markets and the Role of Government*. Toronto: University of Toronto Press.

—1981. Canadian housing policy in the seventies. *Land Economics* 57: 338-52.

—1983. The crisis in rental housing: a Canadian perspective. *Annals of the American Association of Political and Social Sciences* 465: 58-75.

—1984. Household headship rates, household formation, and housing demand in Canada. *Land Economics* 60: 180-8.

Steele, M. 1979. *The Demand for Housing in Canada*. Census Analytical Study. Catalogue No. 99-763. Ottawa: Statistics Canada.

Straszheim, M.R. 1975. *An Econometric Analysis of the Urban Housing Market*. National Bureau of Economic Research. New York: Columbia University Press:

Struyk, R.J. 1976. *Urban Homeownership: The Economic Determinants*. Lexington: D.C. Heath.

Sweet, J.A. 1972. The living arrangements of separated, widowed, and divorced mothers. *Demography* 9: 143-57.

—1984. Components of change in the number of households: 1970-1980. *Demography* 21: 129-40.

Task Force on CMHC. 1979. *Report on Canada Mortgage and Housing Corporation*. Published under the authority of the minister responsible for housing. Ottawa: CMHC.

Thornton, A. 1978. Marital dissolution, remarriage, and childbearing. *Demography* 15: 361-80.

Tsui, A.O. 1982. The family formation process among U.S. marriage cohorts. *Demography* 19: 1-27.

Turchi, B.A. 1975. *The Demand for Children: The Economics of Fertility in the United States*. Cambridge, Mass.: Ballinger Publishing Company.

Uhlenberg, P.R. 1974. Cohort variations in family life cycle experiences of U.S. females. *Journal of Marriage and the Family* 36: 284-92.

Wargon, S.T. 1979a. *Canadian Households and Families: Recent Demographic Trends*. Census Analytical Study. Catalogue 99-753. Ottawa: Statistics Canada.

—1979b. *Children in Canadian Families*. Catalogue 98-810. Ottawa: Statistics Canada.

Index